Copernicus Books

Sparking Curiosity and Explaining the World

Drawing inspiration from their Renaissance namesake, Copernicus books revolve around scientific curiosity and discovery. Authored by experts from around the world, our books strive to break down barriers and make scientific knowledge more accessible to the public, tackling modern concepts and technologies in a nontechnical and engaging way. Copernicus books are always written with the lay reader in mind, offering introductory forays into different fields to show how the world of science is transforming our daily lives. From astronomy to medicine, business to biology, you will find herein an enriching collection of literature that answers your questions and inspires you to ask even more.

Volker Ziemann

Beams

The Story of Particle Accelerators and the Science They Discover

 Springer

Volker Ziemann 🆔
Department of Physics and Astronomy
Uppsala University
Uppsala, Sweden

ISSN 2731-8982   ISSN 2731-8990   (electronic)
Copernicus Books
ISBN 978-3-031-51851-5   ISBN 978-3-031-51852-2   (eBook)
https://doi.org/10.1007/978-3-031-51852-2

This Springer imprint is published by the registered company Springer Nature Switzerland AG
The registered company address is: Gewerbestrasse 11, 6330 Cham, Switzerland

Paper in this product is recyclable.

Forces (all of them Bosons)

Electro-magnetic: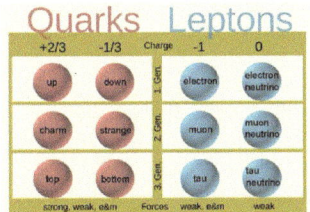
Weak:
Strong: eight x
Gravity: outside standard model

Particles (all of them Fermions)

Quark composites

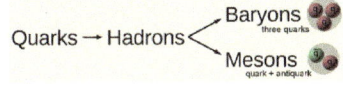

Higgs Boson

Preface

Several years ago, when tying my shoelaces after a visit to our local swimming pool, I overheard two teenagers excitedly discussing the Omega-minus particle. Not believing what I heard, I started talking to them and they confirmed that it was all about particle physics and their enthusiasm about the subject was contagious. Ever since, thinking about this encounter has put a smile on my face and made me optimistic about the future of my field, which is physics, accelerator physics to be more specific. As a day's job I work with particle accelerators, among them were synchrotron radiation sources, a linear collider, the large hadron collider, a free-electron laser, and a small storage ring used for nuclear physics experiments.

Coming from the accelerators I thought it is prudent to complement the many great nonfiction books about particle physics[1] with one covering the role that accelerators played in the game. After all, progress in both fields, particle physics and accelerators, is mutually contingent. New accelerators enable discoveries of new particles and new theories make predictions that motivate building new accelerators to validate them. This give-and-take forms one thread that runs through this book.

And progress is rarely smooth. More often, long tranquil periods of gradual improvements are punctuated by great ideas or great inventions that, all of a sudden, open radically new possibilities. The Internet and the world-wide web are examples from everyday life; they fundamentally changed the way we interact with each other and how we do business. In a similar vein, other great ideas boosted the performance of particle accelerators and others yet, our understanding of the subatomic world. These great ideas form the second thread that runs through this book.

While fleshing out the threads I had my two teenagers in mind as proto-typical readers, or anyone else who might read *New Scientists* or *Scientific American*. The latter was my favorite magazine when I was a (late) teenager with similar interests—yes, I was probably a nerd before the word entered popular culture. At the time, I ordered many reprints of older articles that I recently unearthed in my basement. Among them were gems like Erwin Schrödinger's article *What is Matter?* explaining that sometimes particles behave like, well, particles and sometimes like waves. Rereading many of these old articles I decided to settle for a similar style: no formulas (except $E = mc^2$) and many illustrations to convey ideas and to describe technologies. Basically, I wrote this book for myself as a teenager. You'll be the judge whether that works for you as well, dear reader.

Throughout this book, you'll find references to these reprints, both from Scientific American and from other sources. They provide extra background to selected topics that should be relatively easy to access via public libraries. I collected direct links to most of the articles on https://github.com/volkziem/Beams. Again, often libraries can help to access material behind paywalls. In case you wonder, the Omega-minus and the corresponding magazine article are featured in Chap. 6.

I am grateful to my colleagues and my diligent proof-readers, in particular Stefan Leupold, Roger Ruber, Björn Persson, Ingvar Ziemann, Ellen Matlok-Ziemann, and Elin Bergeås-Kuutmann. All remaining blunders are, of course, mine. If you spot one, don't keep it but rather tell me about it.

Uppsala, Sweden   Volker Ziemann

Note

1. Leon Lederman, *The god particle*, Dell publishing, New York, 1993; Frank Close, *Particle Physics, a very short introduction*, Oxford University Press, Oxford, 2004; Lisa Randall, *Higgs discovery, the power of empty space*, Ecco press, New York, 2013; Harald Fritzsch, *Quarks, the stuff of matter*, Basic books, New York, 1983.

Contents

1

Introduction

A few years ago my publisher set up a web page entitled "What is Physics?" and asked me to contribute an answer. I came up with:

> It's figuring out the world we live in and building the instruments to help with the figuring![1]

One key moment of this figuring process was the press conference at the European Laboratory for Particle Physics (CERN)[2] in the summer of 2012 where the discovery of a new particle was announced.[3] This new particle, known as the *Higgs boson*, was the last missing part of, arguably, the greatest conceptual framework ever conceived by the human mind, the *standard model of elementary particles*.[4] It describes no less than all fundamental interactions and all elementary particles that make up our world.

That this press conference was held at CERN was no coincidence; CERN is the home of one of the largest instruments on earth, the *Large Hadron Collider* (LHC), a 27 km long string of magnets in an underground tunnel close to the city of Geneva in Switzerland. Inside this tunnel, two counter-propagating beams of protons—nuclei of hydrogen atoms—are accelerated very, very close to the speed of light, before they are smashed head-on into each other, creating conditions very similar to those just after the birth of our universe—the Big Bang. Only in these extreme conditions certain elementary particles, among them the Higgs boson, reveal their existence to the large particle-physics detectors ATLAS and CMS.[5]

Getting to the point of observing the Higgs boson required loads of ingenuity, both in "figuring" and in building the instruments. In this book we'll follow the co-evolution of particle physics and of accelerators for the past 150 years.

© The Author(s), under exclusive license to Springer Nature Switzerland AG 2024
V. Ziemann, *Beams*, Copernicus Books,
https://doi.org/10.1007/978-3-031-51852-2_1

Along the way we'll highlight transformative technologies—superconductivity is one example—and "great ideas"—using radio-frequency waves to accelerate particles is another—that made new generations of accelerators possible. Subsequently these new machines satisfied the theoreticians' need for experimental evidence to prove or disprove their wildest ideas. Occasionally their predictions of new particles stimulated building new accelerators; LHC is an example. At other times, experiments at accelerators revealed new particles whose existence flabbergasted the theoreticians and stimulated new theories. An example is the "particle zoo" that grew at an alarming rate in the early 1960s. The dialectic relationship between theories and accelerators will be our guide throughout the book.

Another thread running through this book is the concept of a *scattering experiment*. It is based on smashing things into each other and observing what comes out. This idea extends from Rutherford's experiments a century ago until today's experiments in the LHC. Rutherford and his collaborators directed particles, emitted from the recently discovered radioactive material radium, onto a thin foil of gold (Chap. 3) and observed recoiling particles with a fluorescent screen. In much the same way (Chap. 11) ATLAS and CMS observe whatever comes out after smashing the LHC beams into each other.

Whereas the energy of the radium emissions allowed Rutherford to probe the structure of atoms, much higher-energy probes are needed to explore the inside of atomic nuclei. This is a consequence of the quantum mechanical nature of matter that becomes important in the microscopic world. Matter sometimes behaves as a particle and sometimes as a wave,[6] whose wavelength depends on its mass and the speed. The larger their product (mass times speed) is, also referred to as the particles' momentum, the shorter the wavelength is.[7] And a large momentum also means that a particle has a large energy. The trick is thus to accelerate particles to very high energies to make them behave as short-wavelength probes that can "see" structures having sizes comparable to their wavelength (Fig. 1.1). All electron microscopes[8] rely on this quirky feature of the quantum world and so does LHC, albeit with protons.

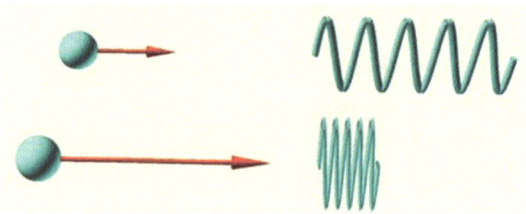

Fig. 1.1 The corresponding wavelength of a faster particle (front) is shorter

As particles reach higher and higher speeds and energies they enter the realm of Einstein's[9] theory of relativity. It predicts that the speed of particles can only increase up to the speed of light as the ultimate speed limit. But where does the energy go when it cannot increase the speed of particles? Einstein found that instead their mass m increases; so energy is going into the mass. As a consequence, even at rest, mass represents some amount of energy E, given by Einstein's famous equation $E = mc^2$ where c denotes the speed of light. Throughout, we will use this equivalency between mass and energy to specify the mass of particles. Writing numerical values for specific particles, however, quickly becomes tedious, because the mass of particles is very small and the speed of light is very large. The large number of zeros that need to be written becomes rather cumbersome. This calls for a more efficient notation.

Scientific and Prefix Notation

Light reaches the moon in a little over a second because it travels at the incredibly high speed of close to 300 000 000 meters per second. Finding the equivalent energy of some mass with $E = mc^2$ even requires squaring this large number, which results in a number with sixteen zeros. Since handling these large numbers becomes rather tedious, a shorthand notation was invented. Noting that the initial "3" is followed by eight zeros, we write this number as 3.0×10^8. Converting between these two notation is easy; 10^8 just instructs us to move the decimal point in 3.0 eight places to the right. This simple rule also works for negative numbers in the exponent. We only have to move the decimal point to the left; 2.0×10^{-3} thus becomes 0.002. Using this "move decimal point" rule makes it obvious that we can add the exponents of numbers; $3 \times 10^8 \times 3 \times 10^8$ thus simplifies to 9×10^{16}. This scientific notation will save me from writing and you from reading a lot of zeros.

And there is another space saver in store. We all know that the prefix "k" for *kilo* translates into "one thousand" or 10^3; a kilometer is one thousand meters. Instead of writing "kk", the prefix "M" for *mega* translates to one million or 10^6. By convention, for every factor of 1000 a new prefix is introduced: "G" for *giga* or 10^9, "T" for *tera* or 10^{12}, and "P" for *peta* or 10^{15}.

In much the same way small numbers are described by prefixes. The letter "m" for *milli* denotes "one thousandth" or 10^{-3}; there are one thousand millimeters in a meter. Again, every factor of 1000 smaller numbers are described by a new prefix: "μ" for *micro* or 10^{-6}, "n" for *nano* or 10^{-9}, "p" for *pico* or 10^{-12}, and "f" for *femto* or 10^{-15}. There are additional prefixes beyond those introduced here, but we will not need them. We will, however, liberally use both the prefix and scientific notation throughout.

Unification

But let us get back to physics and the people that work in the field. One principle that guides the thinking of many physicists is a desire to *explain as many phenomena as possible based on as few assumptions as possible.* This is what physicists call *unification.* The "many phenomena" come of course from experiments, often done with accelerators, whose results are distilled into an underlying framework that we then call a "theory." Much of the theory-building process involves the *classification* of observations in order to identify systematic traits. In high-energy physics that will concern us throughout this book these traits are common features of groups of elementary particles. Once such a feature is found, we give it a name such as "strangeness" (Chap. 5) or "charm" (Chap. 8). In other words, we invent a classification scheme. Only later, a mathematical framework emerges to explain this classification and to predict new features that are hopefully verified in new experiments. In the physics of the subatomic world this cycle of experiments, classification, and theoretical unification repeats itself over and over again leading to the current state of the art. This is the standard model of elementary particles with the Higgs boson announced at the press conference at CERN.

This desire to explain the world, however, started much earlier. Already in the seventeenth century, based on Tycho Brahe's celestial observations, Johannes Kepler[10] deduced laws, today called "Kepler's laws," for the motion of planets around the sun. At about the same time Galileo Galilei[11] found laws that govern falling bodies on earth. About a century later, Isaac Newton[12] "unified heaven and earth" by formulating a theory that encompasses things happening on earth and in the sky. He even derived Kepler's laws from his "unified theory of gravity." In the beginning 20th century Einstein went one step further and, in his general theory of relativity, extended the scope to even describe different universes. That's some unification!

In much the same way electricity, first explored by Benjamin Franklin and Alessandro Volta,[13] and magnets, explored by Ampere, Oersted, and Faraday,[14] were unified by James-Clerk Maxwell.[15] He found equations, today bearing his name, that placed electricity and magnetic phenomena in a common framework. He encapsulated all observations and empirical laws found by his predecessors into just four equations—as we write them today. They describe how the central quantities of his theory, the electric and magnetic fields, behave. And these equations could do so much more than just explain previous observations. They predicted new phenomena that were soon experimentally found: electro-magnetic waves are an example. Today we use them to communicate via smartphones and to accelerate particles in accelerators. Even the design of magnets that guide particles in accelerators is based on Maxwell's equations.

Gravity and electromagnetism are two of the four fundamental forces found in nature. The inside of this book's front cover shows a brief summary of our current understanding of the four natural forces and particles that may serve as a reminder and a road-map on our journey through history. The two other forces are known as the "weak" force (or weak interaction) and the "strong" nuclear force (or strong interaction). They only make themselves known in the subatomic or nuclear realm. The former plays a central role in the spontaneous decay of particles that characterizes *radioactivity*. The strong nuclear force, on the other hand, is responsible for the stability of atomic nuclei. The existence of both interactions was only realized in the 1930s. But from then on they play a key role in the development of the standard model of particle physics. Check out the time line of the co-evolving history of accelerators and the physics of elementary particles in the back of the book. You'll see that this time line started a bit earlier. As a matter of fact, accelerator prehistory started only a few years after Maxwell had published his theory.

Notes

1. The complete answer is actually a bit longer. Here it is: "Physics? It's figuring out the world we live in and building the instruments to help with the figuring! Let me elaborate: physics strives to understand the inner workings of natural and technical phenomena and explains complex processes through a few basic assumptions. But validating this reasoning involves experiments that often require instruments beyond the state of the art: for example, telescopes that scan the universe to detect radio and gravitational waves, and cutting-edge experiments at particle accelerators, which are used to examine the microscopic world."
2. CERN is the acronym for *Conseil Européen pour la Recherche Nucléaire*, the original French name for the lab.
3. You can watch the press conference on https://www.youtube.com/watch?v=AzX0dwbY4Yk and the press release is available from https://home.cern/news/press-release/cern/cern-experiments-observe-particle-consistent-long-sought-higgs-boson.
4. There are strong indications that there is much more, but we postpone that discussion to the epilogue.
5. Peter Higgs (b. 1929) and Francois Englert (b. 1932) received the Nobel Prize in Physics in 2013 "for the theoretical discovery of a mechanism that contributes to our understanding of the origin of mass of subatomic particles, and which recently

was confirmed through the discovery of the predicted fundamental particle, by the ATLAS and CMS experiments at CERN's Large Hadron Collider."

6. Erwin Schrödinger, *What is Matter?* Scientific American, September 1953, page 52. George Gamow, *The Principle of Uncertainty,* Scientific American, January 1958, page 51.

7. The relationship between the wavelength and the momentum of a particle was first proposed by Louis de Broglie (1892–1987) who received the Nobel Prize in Physics in 1929 "for his discovery of the wave nature of electrons." Importantly, it stimulated Schrödinger to formulate the wave equation that bears Schrödinger's name to describe the motion of electrons in atoms.

8. Electron microscopes focus accelerated electrons to tiny spot sizes (nanometer and even below) and observe electrons or X-rays that are knocked out from the surface. For an overview, see: Thomas Everhart and Thomas Hayes, *The scanning electron microscope,* Scientific American, January 1972, page 54.

9. Albert Einstein (1879–1955) was born in Germany, but emigrated to the US in 1933 after the Nazis came to power in Germany. He fundamentally changed our conception of space and time with his special theory of relativity that he published in 1905, the same year he published an explanation of the photo-electric effect (Chap. 3) in terms of photons. He was awarded the Nobel prize in Physics in 1921 "for his services to theoretical physics, and especially for his discovery of the law of the photoelectric effect." Introducing photons—quanta of light—also made him one of the founding fathers of quantum theory. His life's story is told by many biographers, among them his associate and his secretary: Banesh Hoffmann and Helen Dukas, *Einstein, the Human Side,* Princeton University Press, Princeton, 1972 and Ronald Clark, *Einstein, Life and Times,* Avon books, New York, 1971.

10. Already before the advent of optical telescopes, the Danish astronomer Tycho Brahe (1546–1601) recorded tables with the positions of many celestial objects using a so-called *astrolabe,* a device to accurately determine the positions of stars or planets above the horizon. Based on Brahe's tables, Johannes Kepler (1571–1630) discovered that planets follow elliptical orbits and that the size of the ellipse is related to time a planet needs to complete one revolution. At the same time of taking a large step towards our modern attitude toward sciences Kepler had one foot in the medieval traditions of his day. He prepared horoscopes for dignitaries and had to see his aunt burning at the stake and protect his mother from the same fate. For a short account of his life's story, see: Florin Cajori, *Johannes Kepler,* The Scientific Monthly, May 1930, page 385.

11. Transcending the scholastic, scripture-based science of his day with practical experiments, Galileo Galilei (1564–1642) found that heavier objects fall down as fast as lighter objects and that the periodicity of a pendulum only depends on its length, but not on its weight. This makes him the father of modern scientific methodology. He even applied it to the heavens, by using optical telescopes to observe the moons of Jupiter.

12. Isaac Newton (1642–1727) formulated basic laws that govern the motion under the influence of forces. Moreover, he developed a new field of mathematics, dif-

ferential calculus, that, together with his laws, made it possible to describe not only Galileo's, but also Kepler's observations. The publication of his "Principia Mathematica" sets the starting point of what today is known as the field of "classical mechanics". His life's story is told in: Bernard Cohen, *Isaac Newton,* Scientific American, December 1955, page 73.

13. Benjamin Franklin (1706–1790) invented the lightning rod to give lightnings a safe route to ground without igniting a building on the way. He thus identified electricity as a flowing entity that occurs in two polarities. Alessandro Volta (1745–1827) constructed the first batteries and gave all scientists since a controlled source of electricity for experiments. Read about his life and discoveries in: Giorgio de Santillana, *Alessandro Volta,* Scientific American, January 1965, page 82.

14. Hans-Christian Oersted (1777–1851) discovered that electric currents create magnetic fields that turn the needle of a compass, an observation that incited Andre-Marie Ampere (1775–1836) to study the relationship of electric and magnetic phenomena further. In particular, he noted that two current-carrying wires attract each other and that coiled-up wires, he called such devices *solenoids,* generate much higher magnetic fields than individual wires. Michael Faraday (1791–1867) discovered that time-varying magnetic fields produce electric currents, a process he named *induction,* and used it to construct electric motors and generators. Importantly, Faraday introduced the concept of electric and magnetic *field lines.* Being asked by a politician about the usefulness of electricity, he purportedly answered "One day, sir, you will tax it." How true! Faraday's remarkable development from bookbinder apprentice to Fellow of the Royal Society is described in: Herbert Kondo, *Michael Faraday,* Scientific American, October 1953, page 90.

15. James-Clerk Maxwell (1831–1879) condensed all experimental evidence regarding electric and magnetic phenomena into a concise set of four equations (based on Newton's differential calculus). Apart from predicting the existence of radio waves, he was able to derive their speed of propagation from a few fundamental constants. Remarkably, his equations already incorporate concepts from Einstein's relativity. As a matter of fact, Einstein's first paper on relativity actually deals with this aspect. Maxwell's life and work are told in: James Newman, *James Clerk Maxwell,* Scientific American, June 1955, page 58.

2

Accelerator Prehistory

Around 1875, only a few years after Maxwell had published his theory, William Crookes[1] decided to figure out what electricity really is and built the apparatus, today named *Crookes tube* and shown in Fig. 2.1, to help him with figuring. It is probably the first device that can be called particle accelerator.

In order to be able to observe some electricity-related signal, he covered one end of an evacuated glass tube with luminescent paint that lights up when it is hit by radiation. Moreover, he placed two electrodes inside the tube; one of them formed like a Maltese cross. He then connected these electrodes with cables to the poles of a high-voltage power supply. Nothing showed up on the screen when he connected the positive pole of the power supply to the electrode furthest away from the screen. Reversing the polarity and connecting the negative pole to the distant electrode, the screen lit up with a shadow of the Maltese cross clearly visible. Apparently, "something" came from the distant electrode and traveled towards screen, but was intercepted by the Maltese cross, thus casting its shadow. Since the electrode connected to the negative pole of a power supply was historically called *cathode,* the "something" was called *cathode rays.* The electrode, connected to the positive pole—the Maltese cross—was called *anode;* it seemed to attract these rays.

The high voltage was created by an early relative of the ignition system used in older cars with ignition coil and spark plug. Here a battery powers a coil, called *induction coil,* to generate a magnetic field inside. Rapidly interrupting the current flowing through the coil creates a large surge of voltage, a phenomenon called *induction* that was already described by Maxwell's theory. At the time, like today, creating the high voltages needed to accelerate particles was a formidable challenge and required quite a bit of ingenuity.

V. Ziemann, *Beams,* Copernicus Books,
https://doi.org/10.1007/978-3-031-51852-2_2

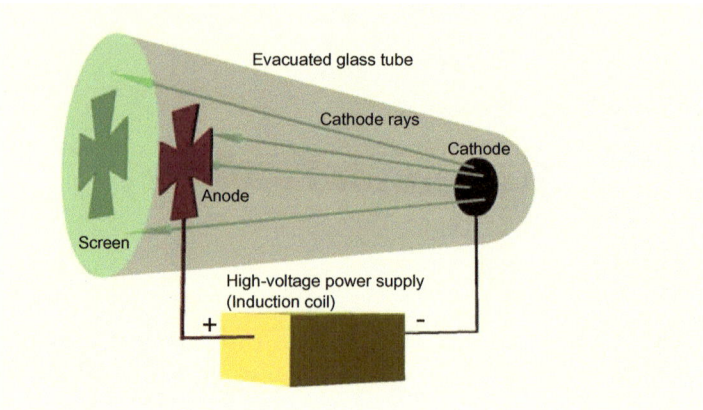

Fig. 2.1 Crookes tube and induction coil based high-voltage source

But let's get back to the rays. Since they move from the negative pole of the high voltage generator to the positive pole, Crookes deduced that these rays are negatively charged. Moreover, he found that they move *parallel* to the electric field between the electrodes. The rays appeared to experience a force that pulls and accelerates them towards the positive pole. Later, Crookes investigated the effect of magnetic fields on the rays and found that they deflect these rays in a direction *perpendicular* to the magnetic field. These two fundamental observations

- electric fields increase energy of charged particles;
- magnetic fields deflect charged particles;

are the basis for practically all accelerators built ever since.

A little over 20 years later in 1897, J. J. Thomson[2] used an apparatus very similar to the Crookes tube. But he added a well-calibrated magnetic field perpendicular to the motion of the cathode rays, which are consequently deflected. For a particular type of particle the deflection angle provides a specific "fingerprint." Heavier particles are deflected less with the same magnetic field and lighter particles are deflected more. Moreover, from the direction of the deflection, either left or right, he deduced that the polarity of the particle is negative. Today, Thomson's measurement device with deflecting magnet is called a *spectrometer* (Fig. 2.2). It is used in practically all particle detectors to identify the type of particles. Henceforth, J. J. Thomson's particle was called *electron*, the first fundamental particle.

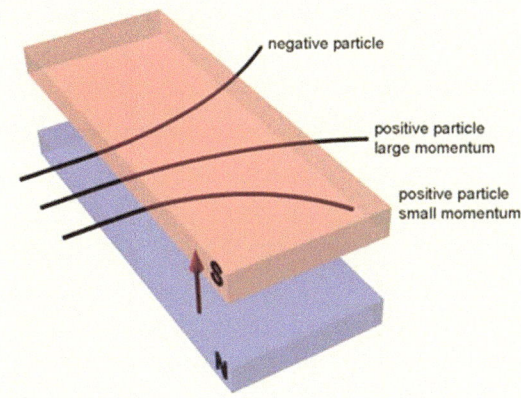

negative particle

positive particle
large momentum

positive particle
small momentum

Fig. 2.2 Spectrometer

After finding out that heating the cathode makes the emission of electrons more reliable the technological spin-off from the Crookes tube was tremendous. The tubes form the base of X-ray tubes that are still used today, more than 120 years after Wilhelm Röntgen's discovery.[3] He observed that cathode rays, impinging on an aluminum window, emitted highly-penetrating radiation that made the bones inside his wife's hand visible. At around the same time Ferdinand Braun[4] improved the technology to steer the cathode rays onto a fluorescent screen. These devices were called cathode-ray tubes and became the base for oscilloscopes, which play a key role in any laboratory to display electric signals. Even more prominent is their use in old-fashioned thick-screen television sets.

A technological spin-off that actually started the field of electronics is the triode, invented in 1906 by Lee de Forest.[5] He added a metallic mesh between cathode and anode. In this way electrons could still pass through the tiny holes and continue towards the anode. But by varying a voltage applied to the mesh, he was able to adjust the intensity of the cathode rays. A small voltage applied to the mesh thus changed the larger electron current traveling from the cathode to the anode. This was in fact the first electronic amplifier. Telephone companies soon used it to amplify weak electric signals, and this made long-distance telephone calls possible. Changing the mesh voltage by a large amount even turns the electron beam on or off, which is just what an electronic switch does. Importantly, it could be operated much more quickly than mechanical switches. These tubes with cathode, mesh, and anode were called triode and are still used today in some applications. Being able to rapidly switch currents made it possible to build fast-switching electronic circuits, so-called *oscillators.* They played a key role for the emerging radio broadcasts but also for accelerators.

The voltage between cathode and anode determines the energy of the electrons. Let us therefore briefly establish the relationship between voltages and energies.

Electron Volts

Crookes managed to reach voltages U in the range of 50 000 V between cathode and anode. This accelerated the electrons to an energy of eU, where e is the charge of the electron.[6] We can compare this with the energy $E = mc^2$ that is equivalent to its mass[7] and find that eU is about 10% of mc^2. At this voltage, their speed turned out to be a little more than 10^8 m/s.

Using the conventional units, based on kilogram, meter, second, and Ampere, the numerical values of energies and speed turn out to be very large or very small. This calls for a better system of units. Scientists use the equivalent voltage $\hat{U} = E/e$ to describe all energies E that appear in calculations. The units of $e\hat{U}$ are called electron-volts with the abbreviation eV. Applying this to the mass of the electron, we find that $e\hat{U} = mc^2 = 511$ keV, where we immediately use the "k" to denote a factor of 1000. The mass m of the electron is thus 511 keV$/c^2$. By convention, the c^2 is absorbed into the units, without its numerical value explicitly entering the calculations where most factors of c normally cancel anyway. We will use the eV throughout this book to denote energies and eV$/c^2$ to denote masses. Using these units, the mass of protons[8] turns out to be 938 MeV$/c^2$.

Expressed in meters per second, the accelerated particles very quickly reach speeds that require many zeros to write. Since the speed of light is the ultimate speed limit for particles it makes sense to denote speeds as percentage of the speed of light. In this way it becomes immediately clear when the peculiar effects of Einstein's relativity become important. This is typically the case at speeds larger than a few percent of the speed of light.

Reaching much larger voltages and therefore energies was only possible by the late 1920s. Until then, the experimenters had to rely on particle sources that emit high-energy particles all by themselves. Luckily, nature had them in store.

Notes

1. Apart from his work on vacuum tubes, William Crookes (1832–1919) was researching the emission spectra, the optical fingerprints, of heated materials. For a brief account of his achievements see: John Sullivan, *Sir William Crookes*, Scientific American, April 1919, page 396.

2. Joseph John Thomson (1856–1940) discovered the electron and thereby introduced the mass-to-charge ratio as a means to uniquely identify particles, even positively charged ones. In this way he found small differences in the mass of particles that otherwise appear chemically identical, so called *isotopes*. Later the difference was tracked to numbers of neutrons in the different nuclei. He received the Nobel Prize in Physics in 1906 "for his theoretical and experimental investigations on the conduction of electricity by gases." Notably, a large number of later Nobel laureates started their career in the Thomson's Cavendish laboratory in Cambridge, England. A brief outline of his life and achievements can be found in the *Obituary Notices of Fellows of the Royal Society,* December 1941, page 586.

3. Wilhelm Conrad Röntgen (1845–1923) received the first ever Nobel Prize in Physics in 1901 "in recognition of the extraordinary services he has rendered by the discovery of the remarkable rays subsequently named after him." Henceforth it was used to look into human bodies to diagnose broken bones or other maladies. The story of the discovery is told in: Graham Farmelo, *The Discovery of X-rays*, Scientific American, November 1995, page 86. For a vivid narrative of this discovery (and eleven others) see: Suzie Sheehy, *The Matter of Everything*, Bloomsbury Publishing, London, 2022.

4. Ferdinand Braun (1850–1918) developed the first cathode ray tube and used it to visualize electrical signals in what we today call an oscilloscope. He later contributed to the development of the emerging wireless technology for communication. He received the Nobel Prize in Physics in 1909 for "contributions to the development of wireless telegraphy." For more background see: George Shiers, *Ferdinand Braun and the Cathode Ray Tube*, Scientific American, March 1974, page 92.

5. The American Lee de Forest (1873–1961) invented the triode vacuum tube, which started the field of electronics. He went on and used this and his other many inventions to improve the technology of wireless communication.

6. This is one unit of a fundamental charge and has the numerical value $e = 1.602 \times 10^{-19}$ As.

7. The mass of the electron is $m = 9.1 \times 10^{-31}$ kg.

8. The mass of a proton is 1.67×10^{-27} kg.

3

Nature's Accelerators

The cathode-ray tube could only produce electrons with moderate energies, but fortunately nature provides natural sources of radiation with higher energies that turned out to be tremendously useful for exploring the microscopic world. Just before the turn of the century Henri Becquerel[1] found that some elements emitted a new type of penetrating radiation all by themselves. He called this phenomenon *radioactivity*.

Radioactivity

Becquerel had placed well-shielded photographic plates in a drawer and found them inexplicably darkened. Something must have exposed them. When he tried to figure out what did that, the culprit turned out to be some uranium salt that he had left in the same drawer. Somehow the uranium emitted radiation all by itself; no stimulus was needed. In the ensuing period of intense experiments, Becquerel, Marie Curie,[2] and her husband Pierre were able to extract several radioactive substances; in particular *radium* turned out to be a powerful source of the new radiation.

Just a few years later, Ernest Rutherford[3] carefully analyzed the radiation that was emitted from different radioactive elements. When passing the radiation through a magnetic field, one type was deflected to the right—he called it alpha radiation—one type was deflected to the left—that's beta radiation—and one type went straight ahead—that's gamma[4] radiation. From the deflection, he deduced that alpha-radiation consists of rather heavy, positively charged particles. The beta-radiation consists of lighter negatively charged particles,

© The Author(s), under exclusive license to Springer Nature Switzerland AG 2024
V. Ziemann, *Beams*, Copernicus Books,
https://doi.org/10.1007/978-3-031-51852-2_3

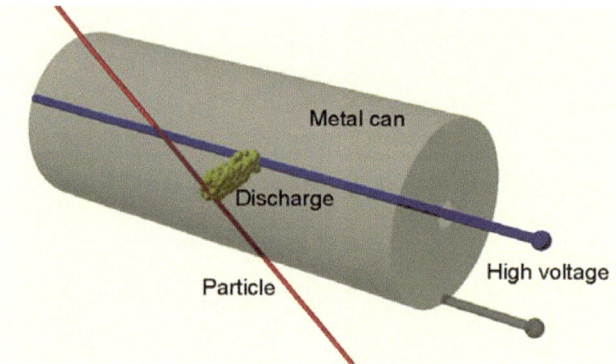

Fig. 3.1 Geiger counter

which turned out to be the same as Thomson's electrons (Chap. 2). The third type behaved in much the same way as Röntgen's X-rays.

In particular the alpha-radiation turned out to be a marvelous source of particles to probe other materials, but first Rutherford had to figure out what the alphas are. Rutherford therefore placed a radioactive source of alphas next to an evacuated glass tube and waited. After some time, he ignited an electric arc inside the tube and observed light with colors that are characteristic of helium gas. This led him to the conclusion that alphas are closely related to helium atoms; only that helium gas is neutral whereas alpha particles are charged.

Knowing what the alphas are, Rutherford used them to irradiate other substances. With only nitrogen gas inside the tube, he noticed that positively charged particles, later identified as *protons*, were ejected with high speed. Moreover, after waiting some time, he found oxygen in the tube. It appeared as if alphas converted some of the nitrogen into oxygen. This conversion of one element into another was the dream of medieval alchemists. Today we know that it involves a nuclear reaction which, in modern parlance, is called *transmutation.*

In order to perform careful experiments with the alphas, he also had to detect them reliably. Rutherford's collaborator Hans Geiger[5] came up with the idea to use a gas-filled metal tube with a wire at high voltage in its center. If an alpha particle passes through the gas, it knocks out electrons from the gas that are accelerated by the high-voltage and thus causes a surge of electric current between the wire and the outer cylinder, which are detected and counted. This device was ever since called a *Geiger counter* (Fig. 3.1). His second idea was to use zinc-sulfide as a material that, if hit by alpha particles, emits a flash of light.

Great Idea: Scattering Experiment

The zinc-sulfide scintillators were soon put to good use when Rutherford and collaborators exposed a thin gold foil to alpha-radiation and used the scintillators to observe the direction of the deflected alphas. They used a microscope with a small piece of zinc-sulfide pasted to the lens and moved it around the gold foil to observe deflected particles from different angles; Fig. 3.2 illustrates the experiment. To their utter surprise, they observed alphas at very large deflection angles, some of them even traveled back towards the source. Rutherford realized that in order to reverse the direction of fast-moving projectiles, very large forces had to be involved. Since no other fundamental forces were known at the time, he assumed they were of electric nature. To account for the experiment, he had to assume that the charged alpha particles approach some other charge to a distance about a hundred thousand times smaller than the size of an atom. In his view the atom consisted of a very dense and positively charged nucleus in the center with electrons circling around it, analogous to planets circling around the sun in the center of the solar system. As shown on the small insert in Fig. 3.2, most alphas miss the tiny nucleus, but if one is headed straight at the nucleus, it can be deflected by a large angle.

Shooting particles at a target and observing what comes out makes this experiment the prototypical *scattering experiment*, which forms the base of practically all modern experiments in nuclear and high-energy physics. Moreover, Rutherford's experiment provided the foundation for the quantum revolution during the 1920s and 30s. Let's see how that got under way.

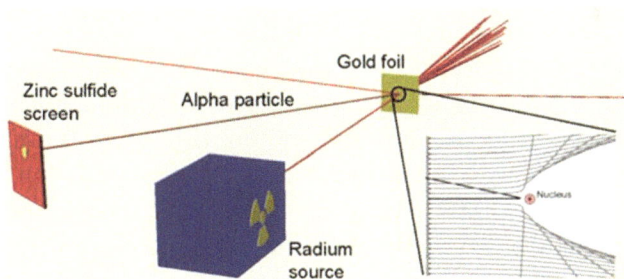

Fig. 3.2 Rutherford's scattering experiment

Rutherford-Bohr Model of the Atom

Soon it was realized that Rutherford's view of a positively charged nucleus with electrons circling around it had a fundamental flaw, because Maxwell's theory predicts that circling charges emit radio waves. This would cause the electrons to lose all their energy within a very short time. But atoms are stable and do not emit waves all by themselves; something must prevent them from doing that.

Finding a way to bypass the problem—"explain" is too strong a word for an ad-hoc assumption, even if it turns out to be very successful—is one of Niels Bohr's[6] greatest contributions to modern physics. He postulated that the electrons can only move on certain orbits around the nucleus and can only lose energy when jumping from one such orbit to another. This postulate did not come from nowhere. He was aware of Max Planck's[7] earlier work on the emission of radiation where Planck had to assume that energy could only be emitted in discrete packets that he called *quanta*.[8] Additionally, shining light on certain surfaces knocked out electrons whose speed depended on the color of the light. Higher light intensities gave more electrons, but they all had the same energy.[9] Albert Einstein explained this observation by assuming that light of a certain color is packaged into quanta, he called them *photons*, that carry a fixed amount of energy. When Bohr worked out the consequences of his postulate, he found that the wavelength (color) of the radiation that his atoms could emit, agreed with the wavelengths observed in experiments.

Bohr's postulate of stable orbits introduced the concept of the *state* of the system—here the electron—and transitions between different states. He labeled the permitted states by *quantum numbers*. There is nothing special about them. They are more like the house number labeling the state that the system presently occupies. The electron orbit closest to the nucleus has the lowest quantum number and the electron is most tightly bound to the nucleus and has therefore the lowest energy. A photon with the right energy can lift the electron to an orbit with a higher quantum number and also higher energy. If the electron later jumps back to the lower state, another photon is emitted. Its wavelength is determined by the energy difference between the states. Figure 3.3 illustrates such a *quantum jump*.

It is remarkable that filling up all available states with electrons starting from the lowest one describes the configuration of all chemical elements in the periodic table, but only if we assume that each state contains two electrons. In this way hydrogen, with element number 1, contains only one electron in the lowest state, whereas helium, with element number 2, contains two electrons in the lowest state. Lithium, with element number 3, has two electrons in the

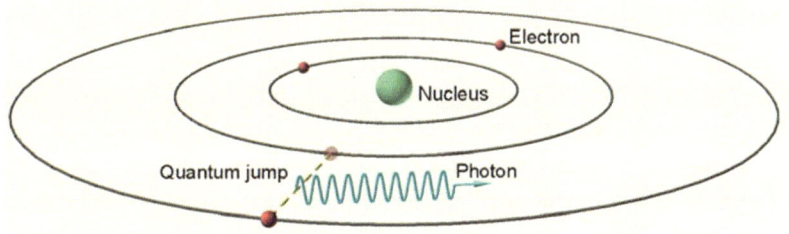

Fig. 3.3 Rutherford-Bohr model of the atom

lowest state and one in the next higher one. Other chemical elements follow the same scheme.

That only two electrons can occupy the same state in an atom was just an observation that gave a consistent picture of how chemical elements are organized. Later, Wolfgang Pauli[10] introduced the concept of *spin* as a property of elementary particles, much like elementary particles carry a charge and have some mass. All electrons turn out to carry spin 1/2, visualized in Fig. 3.4 by electrons spinning like a top with a small arrow pointing along the axis of the top. According to Pauli's reasoning, this arrow can only point in two directions: either up or down, but never in-between. And this gives two electrons a way to occupy the same state in an atom. One electron has its spin pointing upwards and the other downwards.[11] Admittedly, this sounds like black magic, but Pauli's idea provided a classification scheme for elementary particles that was remarkably successful. It not only helped explaining the periodic table but it is also essential to explain the color of light emitted from atoms that are immersed in magnetic fields.[12]

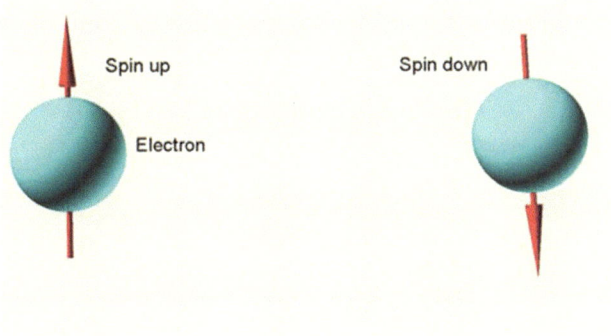

Fig. 3.4 Spin

Neutrons

Despite these successes, the new quantum theory could not explain why atomic nuclei do not explode due to the mutual electric repulsion of all the positively charged protons inside. Moreover, experiments showed that most nuclei were about twice as heavy as the number of protons needed to neutralize the charge of the electrons. Rutherford hypothesized that there are other uncharged particles, he called them *neutrons*, inside a nucleus. He also suspected that they are involved in some, at the time, unknown nuclear force that binds the protons inside the nucleus.[13] As a matter of fact the alpha particles—Rutherford's projectiles—are composed of two protons and two neutrons, which is illustrated in Fig. 3.5. The even simpler combination of a proton and a neutron is called a deuteron.

Only much later in 1932 did James Chadwick[14] discover neutrons by bombarding beryllium with alphas, emitted from the radioactive element polonium. Emerging from the collisions he found a neutral particle that could penetrate lead and also knock out highly energetic protons from a paraffin target[15] that he detected with a Geiger counter. That the new radiation penetrated layers of lead ruled out X-rays; so it had to be a new type of particle, pushing protons makes its mass comparable to that of a proton. Considering that it was knocked out of a beryllium nucleus, it had to be Rutherford's hypothetical neutron.

Once the production of neutrons was understood, they could also be used as probes to irradiate different materials, which is exactly what Enrico Fermi[16] did after he improved his neutron source. This source was now based on radioactive radon irradiating beryllium. Fermi directed the neutrons towards several non-radioactive substances which, he found, became radioactive in the process. Fermi's group noted that paraffin slowed the neutrons down and that gave them more time to interact with the irradiated sample, increasing its radioactivity substantially.

| Proton | Neutron | Deuteron | Alpha |

Fig. 3.5 Proton, neutron, deuteron, and alpha particle. Proton and neutron are jointly referred to as nucleons. Proton and deuteron, having the same charge but different masses, are referred to as isotopes

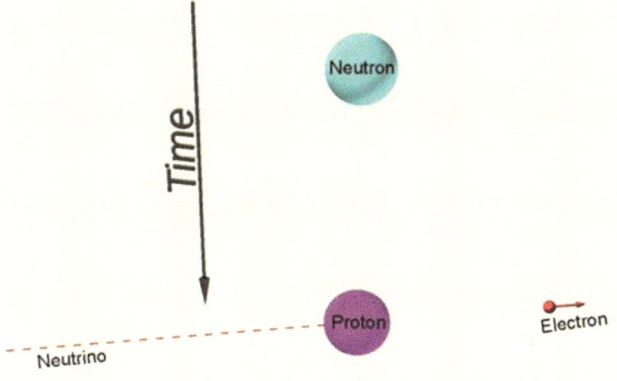

Fig. 3.6 Neutron decay

The neutrons that Fermi produced turned out to spontaneously decay into protons and electrons with only half the neutrons left after a quarter of an hour. The electrons, escaping towards the right in Fig. 3.6, are emitted over a wide range of energies. Moreover, the energies of proton and electron did not add up to the energy (remember $E = mc^2$) of the neutron.[17] Pauli had earlier hypothesized that a third particle, uncharged and undetected, must be involved. It would carry away the missing energy. Fermi gave this particle the name *neutrino*, Italian for "little neutral one," and came up with a theoretical model of decaying neutrons that could explain the experiments. That it takes a quarter of an hour to decay shows that the fundamental force responsible for this radioactive decay of the neutron is very weak; it is the first example of what was later attributed to the third of the fundamental forces, the weak force.

Fermi was not the only one to use neutrons for irradiating materials. Towards the end of the 1930s, Otto Hahn,[18] Fritz Strassmann, and Lise Meitner[19] irradiated uranium with neutrons and painstakingly analyzed the chemical composition of the residues, shown on the right-hand side in Fig. 3.7. To their utter surprise, they found the much lighter element barium. This indicated that they had split the uranium nucleus into two substantially smaller parts, a process called *nuclear fission*. Subsequent theoretical analysis showed that in the process an enormous amount of energy is released and, equally important, a number of neutrons are emitted. Soon Meitner realized that these neutrons could split more uranium atoms, which would give rise to more neutrons and thus lead to a *nuclear chain reaction*. She immediately realized the military threat that this observation posed, if Nazi Germany would build a bomb based on this observation. She therefore passed the news on to international colleagues

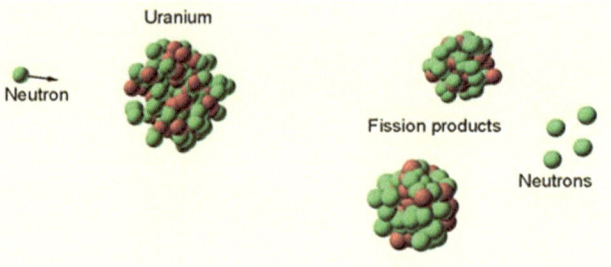

Fig. 3.7   Nuclear fission

and thereby indirectly triggered the development of atomic bombs in the US[20] and, along the way, to nuclear power plants as well.

Radioactivity proved enormously useful for understanding nuclear phenomena, but still, the available energies were limited to a few MeV of the escaping alphas or neutrons. They were useful to coarsely explore nuclei, but in order to "see" more details, particles with higher energies were needed. To find those, the physicists turned towards the heavens and found *cosmic radiation.*

Cosmic Radiation

Fast moving charged particles knock electrons from host atoms and thereby create charged atoms, so-called *ions.* These fast-moving particles are therefore referred to as *ionizing radiation.* Around 1912, Victor Hess[21] took an *electrometer,* a device to measure the intensity of the ionizing radiation, on a high-altitude trip in a balloon, all the way up to 5000 m. He wanted to find out whether the radiation came from the earth below or from the sky above. While ascending he observed the electrometer and wrote down the ionization, which decreased up to a height of about 1000 m before it started to increase again as the balloon climbed to higher altitudes. He even did this at night and during solar eclipses, but always found the same behavior. Apparently, at low altitudes the radiation came from below, but at higher altitudes, the radiation from above dominated—the cosmic radiation, as it was soon called. The left-hand side in Fig. 3.8 illustrates this. The radiation proved a most useful source of high-energy particles.[22]

The electrometer, shown on the right-hand side in Fig. 3.8, was only one device to detect radiation. Stacks of *photographic emulsions*[23] provided another means of detection.[24] Especially after being transported to very high altitudes, the photographic plates occasionally displayed tracks of radiation and even nuclear reactions in which cosmic particles collide with atoms in the plate.

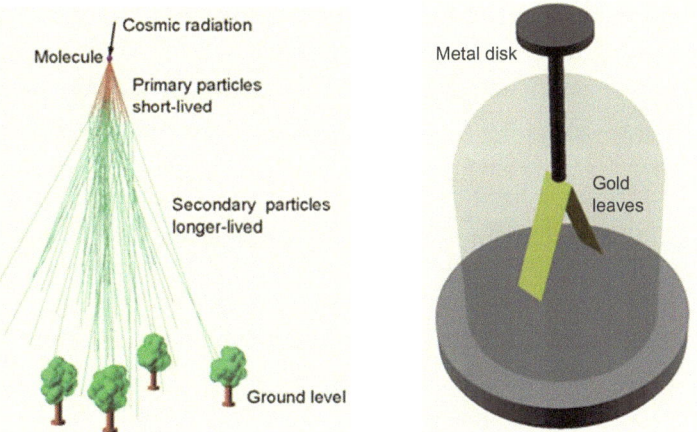

Fig. 3.8 Cosmic radiation (left) and electrometer (right)

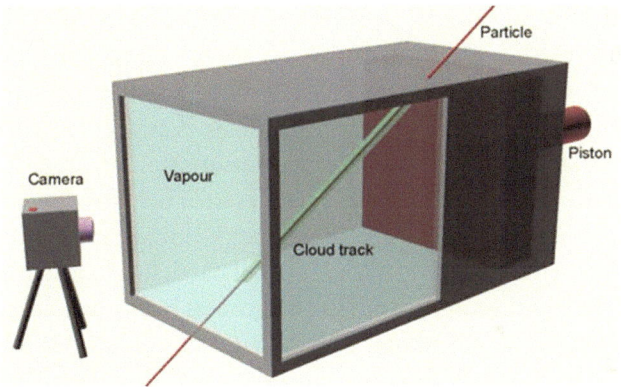

Fig. 3.9 Cloud chamber (track not to scale)

Yet another device to detect charged particles is the *cloud chamber*, invented by Charles Wilson[25] at about the same time Hess did his flights. In a cloud chamber, its basic idea is illustrated in Fig. 3.9, water or alcohol vapor is cooled below its boiling point by rapidly pulling a piston to increase the available volume for the gas. In this state, called super-saturated, the gas wants to turn into a liquid, but can only do so if a nucleation site is present and actually starts the condensation. And this is what passing charged particles do. They ionize the gas and thus initiate the condensation into bubbles, which are then photographed through windows to make the track of the particle visible. Note, however, that the track in Fig. 3.9 is not shown to scale. In reality the width of a track is just a fraction of a millimeter.

Just as research with cosmic radiation took off, the First World War broke out. Only after the war did research resume and consolidated the field by determining characteristic properties of the radiation such as the number of particles per area and their energies. Some particles even turned out to have energies that allowed them to penetrate several lead bricks. Moreover, Geiger counters, placed some distance apart, often triggered simultaneously. This supports the hypothesis that a single high-energy particle, striking an atom in the upper atmosphere, creates a shower of secondary particles that spreads out over a large area, as shown on the left-hand side in Fig. 3.8.

Cosmic rays as particle source and cloud chambers as detectors gave rise to several spectacular results.

Positrons

In the late 1920s Paul Dirac[26] worked out a theory to reconcile the new quantum mechanics, as developed by Werner Heisenberg[27] and Erwin Schrödinger,[28] with Einstein's special relativity. He discovered peculiar solutions to his equations that predicted particles that seemed to have negative energies. In 1928 he came up with a revolutionary interpretation; the energy of these particles is positive, but the particles are antiparticles instead. Such an antiparticle is the mirror image of the corresponding particle in every respect and behaves like a hole into which the normal particle falls and disappears, leaving only a flash of radiation behind. In the process both particle and antiparticle vanish, or, scientifically speaking, they *annihilate* each other. Only four years later in 1932, did Carl Anderson[29] find the curved track on the cosmic-ray-irradiated cloud-chamber image, shown in Fig. 3.10. It had all the appearance of an electron track, but in the magnetic field it curls the "wrong" way. But to know this he had to figure out which way the particle traveled. The 6 mm thick lead plate, shown in the middle of the image, helped him do that, because the particle loses energy in lead and curls more after penetrating the lead. Here the particle apparently comes from below. Once convinced that the particle actually curls in the opposite sense to the more abundant electrons, he announced the discovery of a positively charged electron. This was soon labeled *positron*, the anti-electron Dirac had predicted.

Finding a new elementary particle was a big feat for Anderson, but rather soon he did it again, when he discovered *muons*.

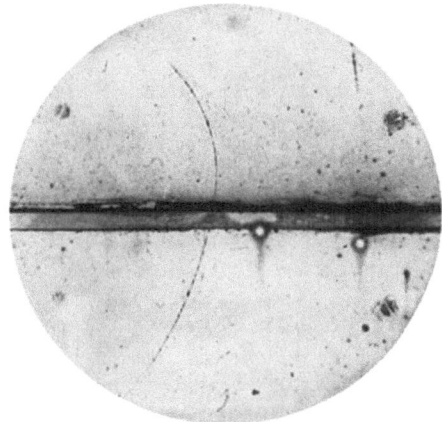

Fig. 3.10 Positron discovery [in the public domain, original *source* Carl Anderson, *The Positive Electron,* Physical Review 43 (1933) 491]

Muons

Muons suffered from a case of mistaken identity when Anderson found particles with a mass of about $100\,\text{MeV}/c^2$ in his cloud chambers. He suspected that this discovery was the nuclear particle Hideki Yukawa[30] had predicted a few years earlier. Yukawa wondered why the positive charges in the nucleus did not fly apart. To counteract the electric repulsion, he postulated a nuclear force that had to be very strong. Moreover, this force acted only over the very small distances inside a nucleus. Based on quantum-mechanical reasoning, he predicted that a particle with a mass of about $100\,\text{MeV}/c^2$ should be involved. That Anderson found a particle with about this mass caused quite some confusion that lasted for quite some time. This dilemma was only cleared up by a group of Italian physicists[31] about a decade later. They found that Anderson's new particle did not participate in nuclear reactions, as it should if it were Yukawa's particle. It therefore had to be a completely new particle that did not fit into any scheme. Isidor Rabi[32] famously asked: who ordered that? Anderson's new particle received the name *muon*.[33] Further experiments showed that, depending on its charge, it behaves exactly like an electron or a positron. The only difference is that it is about 200 times heavier than its lighter pendent. Today, we classify electrons and muons as *leptons,* for "light particles," because they are lighter than most other particles, such as protons or neutrons, which are collectively known as *hadrons.* At this point, this is just an ad-hoc classification that helps us put some order into the, as we shall see, rapidly expanding zoo of new particles.

Pions

But what about Yukawa's nuclear "glue" particle[34]? Just after the Second World War, it was finally found by exposing photographic plates to cosmic radiation at high altitude. Already in 1947, Cecil Powell[35] and his collaborators found a candidate track in a new type of photographic plates that were optimized to study nuclear reactions. Powell had exposed his new plates to cosmic rays on the Pic du Midi in the French Pyrenees at an altitude of almost 3000 m. After taking the plates back to the laboratory, he found a particle with a moderate mass decaying into a second, slightly lighter particle having the mass of a muon. Soon plates were shipped to Bolivia and exposed at an altitude of 5600 m. They revealed ten additional decay chains with the same characteristics. Moreover, the newly discovered heavier particle participated in nuclear reactions as it should if it were Yukawa's particle. Two versions of this particle, which was dubbed *pion*, short for π-meson,[36] appeared, one positive π^+ and one negative π^-.

That pions decay into muons of course contributed to the earlier misinterpretation of Anderson's particle as carrier of the nuclear force. Today we know that the primary pions are created in the upper atmosphere, as shown on the left-hand side in Fig. 3.8. But within a few meters they decay into muons, which have a lifetime of about two millionths of a second, which allows them to travel up to 600 m.[37] That we can observe them reaching the ground, despite their short lifetime, is a consequence of Einstein's special relativity. For us observers at rest, the muon ages slower, often by factors of a thousand. And that is enough for them to reach the surface, where Anderson observed them in his cloud chamber.

The late 1940s were the heydays of cosmic ray research with photographic plates. And they had more to offer.

V-Particles

George Rochester and Clifford Butler had equipped a cloud chamber with Geiger counters, as shown in Fig. 3.11. The Geiger counters were used to trigger the camera once a particle shower was detected and this resulted in a significant increase of "good" pictures. Moreover, using two cameras taking pictures from different positions made it possible to reconstruct the track in three dimensions. On one of their pictures they noted a peculiar V-shaped track that popped up out of nothing. Figure 3.12 illustrates the story that goes like this: first a cosmic-ray particle strikes a nucleus. From the debris emerges a neutral (and therefore invisible) particle, which lives long enough to travel

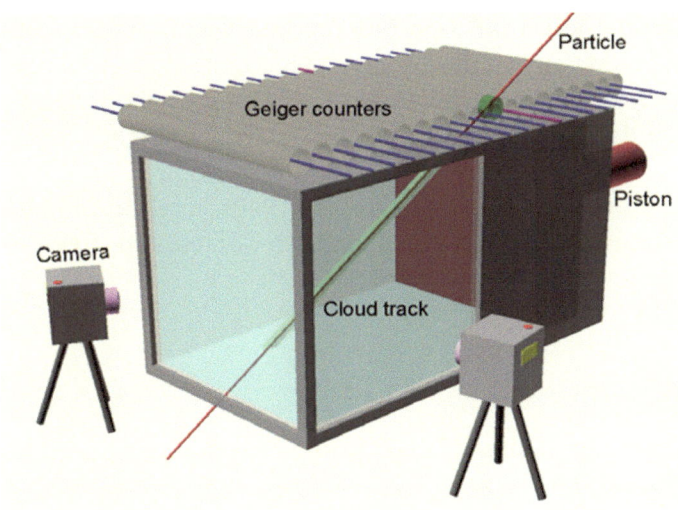

Fig. 3.11 Cloud chamber with Geiger counters as trigger

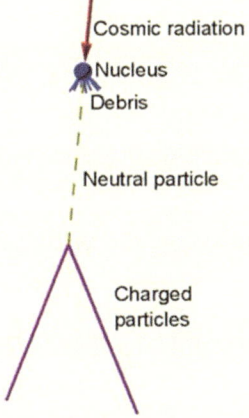

Fig. 3.12 V particles

some distance before it decays into two charged (and therefore visible) particles. Now the question is: why does the neutral particle live so long? It is born in a strong-force collision, which acts only while particles touch each other. But it cannot decay in a strong-force event, because its observed lifetime is too long. This long lifetime suggests that the weak force, normally responsible for radioactivity, is also responsible for the decay of V-particles. It was funny, here we have a particle that is born in a "strong" process, but decays in a "weak" process. That is something to behold.

It took several years until a few more of these V-shaped tracks were found. By determining the energy of their decay products, the mass of the neutral particle was determined to be around $500\,\mathrm{MeV}/c^2$, roughly about half the mass of a proton. In order to understand the nature of these strange new particles, many more such observations were needed and that required particle sources with much higher intensity. Luckily by the late 1940s, accelerators had just reached a level of maturity to become useful in this line of research. To get to that point, however, several great ideas were required.

Notes

1. The French physicist Henri Becquerel (1852–1908) received the Nobel Prize in Physics in 1903 "for his discovery of spontaneous radioactivity."
2. Marie Curie (1867–1934), of Polish descent, jointly with her French husband Pierre Curie (1859–1906), received the Nobel Prize in Physics 1903 "for their joint researches on the radiation phenomena discovered by Professor Henri Becquerel." In 1911 Marie Curie received a second Nobel Prize, this time in Chemistry, for "the discovery of the elements radium and polonium, by the isolation of radium and the study of the nature and compounds of this remarkable element."
3. The New Zeelander Ernest Rutherford (1871–1937) received the Nobel Prize in Chemistry in 1908 "for his investigations into the disintegration of the elements, and the chemistry of radioactive substances." By the time he had discovered the new element radon, introduced the concept of radioactive half-life, and classified the three types of radioactive radiation. After he moved to England, he went on to perform the scattering experiments, mentioned later in this chapter, in which he created the basis for our understanding of atoms. After Thomson's death, Rutherford became director of the Cavendish laboratory, where a number of his collaborators performed work that led to a number of Nobel Prizes, most notably Chadwick, who later discovered the neutron. Rutherford's life is described in the *Obituary Notices of Fellows of the Royal Society*, January 1938, page 394.
4. Alpha, beta, and gamma are the first three letters of the Greek alphabet and are frequently used to label a new concept. The corresponding Greek letters are: α, β, and γ.
5. Hans Geiger (1882–1945) invented the electronic circuit, still carrying his name, to detect and count radiation. Moreover, he was one of the scientists to actually count the flashes in the "Rutherford scattering" experiment. Ernest Marsden (1889–1970) was the other.
6. Niels Bohr (1885–1962) came up with the atomic model of negatively charged electrons revolving on discrete—quantized—orbits around a central positively charged nucleus. He received the Nobel Prize in Physics in 1922 "for his services in the investigation of the structure of atoms and of the radiation emanating from them." Much of quantum theory, and especially the probabilistic interpretation

that was later dubbed "Copenhagen interpretation" was developed by his group and many visitors at his institute in Copenhagen. His life's story is told in: Ruth Moore, *Niels Bohr: The Man, His Science, and the World They Changed*, MIT Press, Cambridge, 1966.

7. While developing a theory to explain the color the radiation emitted by heated objects, Max Planck (1858–1947) showed that the radiated energy can only appear as one discrete packet—a quantum—at a time. This makes him the originator of quantum theory. He was awarded the Nobel Prize in Physics in 1918 "for the services he rendered to the advancement of physics by his discovery of energy quanta."

8. Quanta is the plural of *quantum*, the word that gave the emerging quantum theory its name.

9. Einstein was the first who came up with an explanation of this *photo-electric effect*, for which he received the Nobel Prize in Physics in 1921.

10. Wolfgang Pauli (1900–1958) introduced the idea that many elementary particles behave like spinning tops and called this property *spin*. One of the consequences is the 'Pauli exclusion principle' that prohibits two same-type particles with spin $-1/2$, such as electrons, to occupy the same spot. He received the Nobel Prize in Physics in 1945 "for the discovery of the Exclusion Principle, also called the Pauli principle." Furthermore, he proposed the existence of an unseen particle, later called *neutrino*, to account for the energy that unexplainably disappeared in some radioactive decay processes.

11. George Gamow, *The Exclusion Principle*, Scientific American, July 1959, page 74.

12. This is called the *Zeeman effect*. Immersed in a magnetic field, electrons with opposite spin orientations have slightly different energies, such that quantum jumps cause the emission of photons with slightly different wavelengths.

13. Today we know that the number of neutrons is approximately equal to the number of protons in a nucleus. The variants with different numbers of neutrons are called *isotopes*.

14. James Chadwick (1891–1974) received the Nobel Prize in Physics in 1935 "for the discovery of the neutron." For background information, see: J. Crowther, *And now the Neutron*, Scientific American, August 1932, page 76, and Philip Morrison and Emily Morrison, *The Neutron*, Scientific American, October 1951, page 44.

15. Paraffin has a high contents of hydrogen atoms whose nuclei are protons.

16. Enrico Fermi (1901–1954) received the Nobel Prize in Physics in 1938 "for his demonstrations of the existence of new radioactive elements produced by neutron irradiation, and for his related discovery of nuclear reactions brought about by slow neutrons." After emigrating to the US he built the first nuclear reactor in Chicago and later discovered the Delta resonance (Chap. 5). His wife tells us about his life's story in: Laura Fermi, *Atoms in the Family*, University of Chicago Press, Chicago, 1954.

17. This puzzled the physicists quite a bit and wild theories about energy-nonconservation were launched. For a contemporary account at the time, see:

Jean Harrington, *The Mystery of the Neutrino,* Scientific American, May 1936, page 256.

18. Otto Hahn (1879–1868) received the Nobel Prize in Chemistry in 1944 "for his discovery of the fission of heavy nuclei."

19. Lise Meitner (1878–1968) provided the interpretation and potential dangers inherent in Hahn's and Strassmann's findings.

20. See footnote 22 in Chap. 4 for a brief discussion and references.

21. Victor Hess (1883–1964) received the Nobel Prize in Physics in 1936 "for his discovery of cosmic radiation."

22. For an early historic perspective see: P. Blackett, *Cosmic Radiation,* Scientific American, November 1938, page 246.

23. Photographic emulsions are thin films of silver-bromide or other silver-halide crystals coating a substrate. A passing charged particle induces a chemical change in the crystals, just like light exposes a photographic film in a camera, that makes the particle track visible.

24. Herman Yagoda, *The Tracks of Nuclear Particles,* Scientific American, May 1956, page 40. See also A. Herz and W. Lock, *Nuclear Emulsions,* CERN Courier, May 1966, online available from https://cerncourier.com/a/nuclear-emulsions.

25. Charles Wilson (1869–1959) received the Nobel Prize in Physics in 1927 "for his method of making the paths of electrically charged particles visible by condensation of vapour."

26. In his work to generalize Schrödinger's theory and incorporate Einstein's relativity, Paul Dirac (1902–1984) came up with an equation that describes both Pauli's spin and particles with negative energies. He received the Nobel Prize in Physics in 1933 "for the discovery of new productive forms of atomic theory." Dirac realized that in some subatomic reactions the number of particles that are involved changes. An example is the annihilation of an electron and a positron giving rise to a photon. These processes required another extension of the existing theories. He went on to create the first candidate for such a so-called *field theory* that was later dubbed *quantum electrodynamics* (QED). His life's story is told in: Graham Farmelo, *The Strangest Man,* Faber and Faber, London, 2009.

27. Werner Heisenberg (1901–1976) introduced a new method to calculate quantum properties of particles. One of the consequences is the "uncertainty principle" that prevents two related physical quantities, for example the position and the speed of a particle, simultaneously to be determined with infinite precision. He received the Nobel Prize in Physics in 1932 "for the creation of quantum mechanics, the application of which has, inter alia, led to the discovery of the allotropic forms of hydrogen."

28. Erwin Schrödinger (1887–1961) developed an equation, today called "Schrödinger equation," to calculate quantum properties of nature, such as the energy of photons emitted from atoms, with high accuracy. He received the Nobel Prize in Physics in 1933 "for the discovery of new productive forms of atomic theory." His life's story is told in: John Gribbin, *Erwin Schrödinger and the Quantum Revolution,* Transworld books, London, 2013.

29. Carl Anderson (1905–1991) received the Nobel Prize in Physics in 1936 "for his discovery of the positron." A few years later he even observed muons for the first time.

30. After pions were experimentally discovered, Hideki Yukawa (1907–1981) received the Nobel Prize in Physics in 1949 "for his prediction of the existence of mesons on the basis of theoretical work on nuclear forces."

31. Having moved to a school near the Vatican to avoid the bombs that fell on Rome during the war, Marcello Conversi, Ettore Pancini, and Oreste Piccione constructed a crude magnetic lens to separate positive and negative mesons from cosmic-ray showers. They had expected that the negative mesons were captured in an absorber. Instead, they found that the mesons decayed into electrons after a few microseconds, which they could detect with their ingenious electronic circuitry. Yukawa's pions should have been captured in the absorber; therefore the arriving mesons must be something else—muons.

32. Isidor Rabi (1998–1988) received the Nobel Prize in Physics in 1944 "for his resonance method for recording the magnetic properties of atomic nuclei."

33. For a historic view see: Sheldon Penman, *The Muon*, Scientific American, July 1961, page 46.

34. For an early view see: Robert Marshak, *Pions*, Scientific American, January 1957, page 84.

35. Cecil Powell (1903–1969) received the Nobel Prize in Physics in 1950 "for his development of the photographic method of studying nuclear processes and his discoveries regarding mesons made with this method."

36. The word "meson" derives from the Greek word "meso" for "middle," because the mass lies between that of electrons and protons. Historically, even muons were called μ-mesons, but later it was realized that muons belong to a different class of particles, the leptons.

37. The muon has an average lifetime of 2.2 μs. If the muon travels close to the speed of light $c = 3 \times 10^8$ m/s, it covers the distance 3×10^8 m/s $\times 2.2 \times 10^{-6}$ s $= 660$ m in its own system of reference. We non-moving observers on the ground see the muon aging much slower—just as the fast-moving twin in the spaceship ages much slower according to the twin-paradox of special relativity. And this gives the muons enough time to reach the surface.

4

The Echo of Rutherford's Call

In his presidential address to the Royal Society[1] in 1927, Rutherford pointed out the need for controllable radiation sources by asking for a *"copious supply of atoms and electrons that have an individual energy far transcending that of the alpha and beta particles."* In other words, he dreamt of accelerators, and more importantly, he encouraged his collaborators to think about ways to build such devices. John Cockroft and Ernest Walton[2] had listened.

Cockroft–Walton

Neutrons, protons, and their aggregates, alphas, are very strongly bound inside nuclei. In order to escape they need energies around 10 MeV to overcome this barrier. But still, some radioactive nuclei happily eject alpha particles. How does that happen? George Gamow[3] found the solution. Alpha particles, trapped inside the nucleus, can take a shortcut *through* the barrier, rather than climbing over it. Taking this shortcut is possible thanks to another quantum-mechanical quirk called *tunneling*. It predicts that particles, with a very small probability, can spontaneously show up on the other side of a barrier, too high to overcome otherwise. Tunneling works both ways. It provides an escape route for particles from the inside of the nucleus, but it also allows particles that move towards the nucleus to get in. Moreover, the height of the barrier depends to a large extent on the electric repulsion between the incident positively-charged particles and all the other particles that make up the nucleus. Therefore, it should be easier to get inside light nuclei that are composed of only a few protons and neutrons.

© The Author(s), under exclusive license to Springer Nature Switzerland AG 2024
V. Ziemann, *Beams*, Copernicus Books,
https://doi.org/10.1007/978-3-031-51852-2_4

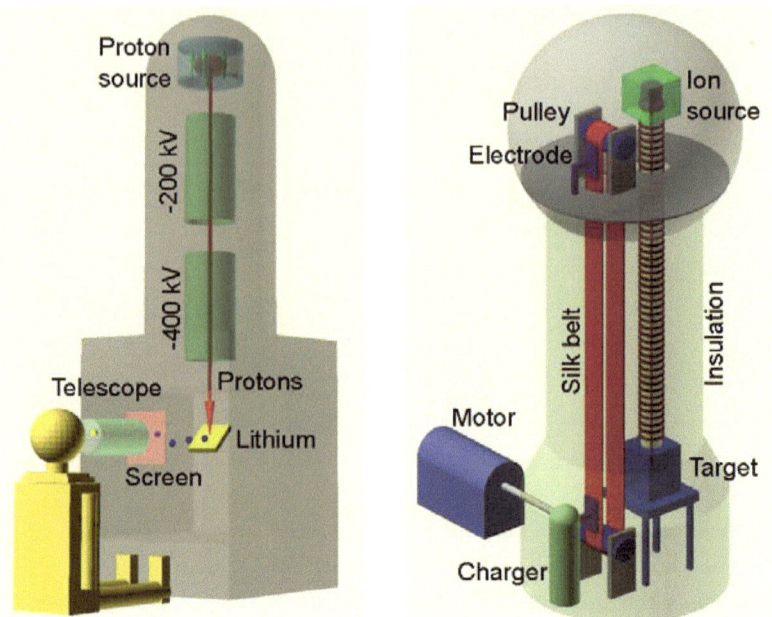

Fig. 4.1 Cockroft–Walton (left) and van de Graaff (right) accelerator

When Cockroft and Walton worked out the math, they found that hydrogen nuclei, protons, with energies of about 600 kV and maybe even less would be sufficient to enter the nucleus of lithium. Lithium is the lightest element on the periodic table after hydrogen and helium. They came up with an electric circuit that connected capacitors and diodes in an ingenious way; a moderately high but time-varying voltage, applied to one side of this circuit, is multiplied and shows up on the other side as a much higher and non-varying voltage.[4] By 1932 their accelerator, shown on the left-hand side in Fig. 4.1, reached 400 kV and was ready for experiments. On the top of their apparatus they produced protons by bombarding hydrogen gas with electrons. The protons are accelerated by the electrodes at high voltage. During their journey, they move inside an evacuated tube that prevents collisions with air molecules. Once the protons reach their maximum energy at the bottom of the tube, they impinge on a lithium target. Occasionally a proton coalesces with a lithium atom and forms a compound that subsequently breaks up into two alpha particles that produce flashes on a screen. Cockroft and Walton sat at the bottom of the high-voltage apparatus staring into a microscope and counting these flashes. This ground-breaking experiment was the first artificially induced nuclear transmutation, here of lithium into two alphas, with an accelerator. Ever since, Cockroft–Walton accelerators are used as the first stage in more powerful accelerators, and they implant ions in semiconductor wafers for microchips (Chap. 13) and solar cells.

Already a few years before, unknown to Cockroft and Walton, a colleague in the US had built an even more powerful accelerator.

Van de Graaff

An air-filled balloon is charged up by rubbing it on one's hair such that it can be pinned to the ceiling. Rubbing transfers charges to the balloon and, since the balloon is non-conducting, they have nowhere to go until they see an escape route at the ceiling, where they slowly leave the balloon. Robert van de Graaff[5] used the same idea when he built his first accelerator in 1929, whose conceptual design is illustrated on the right-hand side in Fig. 4.1. Instead of a balloon, he used a silk belt that constantly rubbed against the lower electrode connected to a charging source, such as a battery. A motor drives the lower pulley, which moves the silk belt and transports the charges up to the upper electrode. This electrode is connected to the large metallic sphere where the charges spread out over its large surface area. Since charges move from the bottom to the top, an electric field builds up in between. Particles created in the ion source at the top are therefore accelerated through an evacuated discharge tube back to the bottom, where they strike a target.

Van de Graaff's first prototype accelerated ions with a voltage of 80 kV while later versions reached several million volts. At some point, however, reaching even higher voltages was limited by uncontrollable discharges. Despite this limitation many van de Graaff accelerators have been built and used. From the time of the Second World War van de Graaff accelerators were used to irradiate targets in order to make them radioactive. Some of the targets subsequently emitted high-energy photons that were shown to break up uranium in a process called photo-fission, which is important for nuclear power generation. Today, van de Graaff accelerators are mainly used for analyzing materials and for dating artefacts (Chap. 13).

The discharges limit the voltages in devices with a single large-voltage acceleration section. Would it not be wonderful, if one could accelerate the particles with lower voltages but do so multiple times? Already in 1924, Gustaf Ising[6] had figured out how.

Great Idea: Radio-Frequency Acceleration

Ising asked himself whether he could repeatedly use the same voltage over and over again and came up with the brilliant idea illustrated in Fig. 4.2, which is still at the core of many of today's linear accelerators. The beam particles are born in an electric discharge on the left-hand side and are accelerated towards the three hollow tubes, labeled 1, 2, and 3. These tubes are excited by an alternating voltage such that the beam is accelerated by the electric field between tubes. It starts by being accelerated towards the first tube. Once inside the tube, the particle is shielded, while the polarity of the tubes reverses. Upon exiting the first tube, the electric field between the first and second tube has the right polarity to accelerate the particle once again. While the particle hides inside the second tube, the polarity reverses again and upon exiting the second tube, it is accelerated towards the third tube. This nifty scheme allows one to reuse a moderate, but oscillating, voltage to repeatedly accelerate particles. Note that the tubes need to become longer as the energy and the speed of the particles increase. Ising never built such an accelerator, but his idea fell on fertile grounds.

Rolf Wideroe,[7] in search of a topic for his Ph.D. dissertation, read Ising's article and built a working prototype with a single drift tube. This earned him his Ph.D. in 1927. Figure 4.3 shows how he turned Ising's idea into reality. On the left-hand side of the evacuated glass container, potassium atoms are ionized through electron bombardment in the particle source. Only particles created at the right moment are accelerated towards the red drift tube, which is excited by a voltage that changes its polarity up to one million times per second. While the potassium ions hide inside the drift tube, the polarity reverses and then the particles are accelerated towards the larger green tube. The high-frequency generator, shown on the bottom, used one of the triodes mentioned in Chap. 2. One of its electrodes, also shown in red, was connected to the red drift tube and its chassis to the other components, colored in green. The generator produced alternating voltages with a maximum value of up to 25 kV to accelerate particles

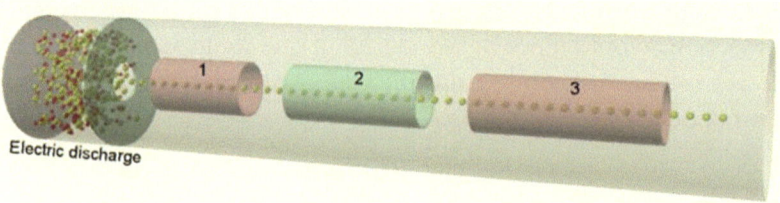

Electric discharge

Fig. 4.2 Ising's idea of a linear accelerator

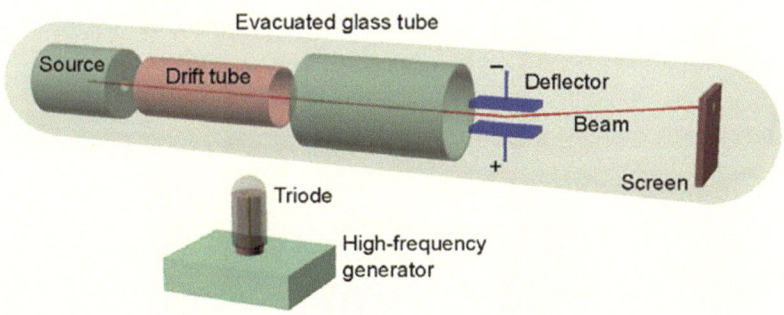

Fig. 4.3   Wideroe's first accelerator

twice, once upon entering and once upon exiting the drift tube. Consequently they reached a maximum energy of 50 keV. Wideroe verified this by deflecting the accelerated particles in the electric field of a deflector, shown in blue, and observed the position of the beam impinging on a phosphor screen. Wideroe published his results in a German journal, left accelerators behind and worked on electrical relays instead.

Wideroe's thesis, however, had a huge impact on the other side of the Atlantic.

Great Idea: Cyclotron

Serendipitously, the university library in Berkeley subscribed to the German journal where Wideroe had published his work. A copy of the journal with Wideroe's article was picked up by Ernest Lawrence,[8] a recently appointed associate professor. He did not know German but immediately figured out the idea behind Wideroe's accelerator from the images in the article. Moreover, he realized that winding up the sequence of tubes into two semi-circular electrodes, called "Dees", and connecting them to a high-frequency oscillator would make the device more compact. He called this new type of accelerator *cyclotron.* Figure 4.4 shows it with some parts pulled forward to make them better visible. Particles are created in a source in the center of the cyclotron and, forced by a strong magnet, spiral outwards. After contemplating his idea and various technical difficulties, he was convinced that it was feasible. But in order to hedge his bets, he decided to charge two students with complementary projects. One was to build a linear and second one a circular accelerator.[9] David Sloan's task was to build the linear accelerator, much like Wideroe's, but with many more

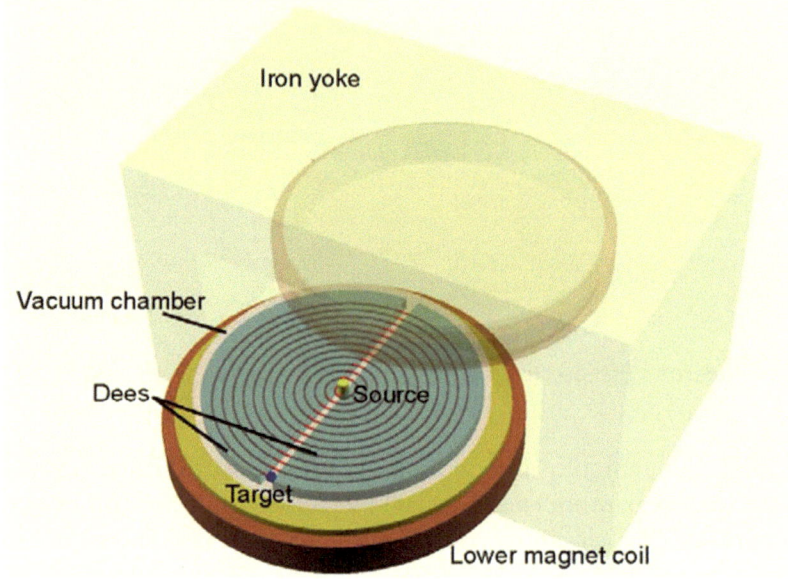

Fig. 4.4 Cyclotron

tubes and acceleration stages. He succeeded magnificently and soon managed to accelerate mercury ions to energies exceeding 1 MeV.

The other student, Stan Livingston,[10] was put in charge of building the first cyclotron based on Lawrence's idea, which proved to be more difficult. In particular, he had to adapt the ion source to produce a beam of H_2^+ molecules.[11] This was necessary to achieve a measurable current at higher energies. He also had to reduce the gas pressure inside the vacuum chamber to minimize collisions of the beam with any remaining gas molecules left inside. Moreover, he had to invent dedicated electrodes to deflect the accelerated beam from its spiral trajectory and direct it to the target, where it was caught and measured. After solving these problems and after borrowing a larger magnet from a neighboring group, he could finally measure a small current of accelerated H_2^+ molecules. They had acquired 80 keV while passing the accelerating gap between the 10 cm or 4 inch semi-circular electrodes—the Dees—82 times.

Lawrence and Livingston immediately started to build a sequence of larger cyclotrons. The next one had a diameter of 11 inches and produced 1.2 MeV protons in 1931. The following one was completed in 1934 and had a diameter of 27 inches. Later the diameter was increased to 37 inches enabling the cyclotron to accelerate deuterons[12] to 8 MeV and alpha particles (Helium nuclei) to 16 MeV. In 1939, an even larger cyclotron, named *the Crocker,* with a diameter of 60 inches, was taken into operation. It typically accelerated deuterons to 16 MeV.

Cancer Therapy

Lawrence's brother John, a medical doctor, joined the lab in 1934. Together, the Lawrence brothers pioneered the use of radiation for biomedical purposes. John substituted artificially produced radio isotopes in biomolecules. This allowed him to trace metabolic pathways in the body, a method used daily in many hospitals even today. He used neutrons that were produced by shooting protons on a beryllium target, just as Chadwick had done, to irradiate cancer cells. He found that they destroyed the tumor more efficiently than X-rays with an equivalent dose. Furthermore, experiments with mice showed that radioactive phosphor-15 had a positive effect on leukemic mice, and later on human patients. But also other types of radiation were tested for irradiations, among them X-rays produced by the accelerator that Lawrence's former student David Sloan had developed.[13]

New Elements and Isotopes

Irradiating samples with high-energy particles created a wealth of new substances with unexpected properties. Since 1938 Martin Kamen and Sam Ruben[14] aimed deuterium beams from the 37-inch cyclotron at a target made of boron, which produced the radioactive isotope carbon-11. Replacing the most common isotope carbon-12 by carbon-11 in carbon-dioxide gas and exposing plants to it, allowed them to follow the route CO_2 takes in the metabolism of plants. An even further-reaching discovery came in 1940 when they directed the deuterium beam from the larger Crocker cyclotron onto a graphite target and found yet another isotope, carbon-14, which has a half-life of about 5730 years. It soon became the key to *radiocarbon dating* of artefacts that was pioneered by Willard Libby.[15] We will return to that topic when we talk about applications of accelerators in Chap. 13.

When Dmitri Mendeleev[16] came up with the periodic table of elements, he noted that he had to leave some empty slots, among them for element number 43. During a visit to Berkeley in 1937, Emilio Segre[17] talked Lawrence into donating a few irradiated components from the cyclotron that he could take back to Italy. Back home, he managed to identify element number 43, which was given the name *technetium*. This new element proved to be tremendously useful for medical diagnosis, because one of its isotopes emits photons of a characteristic energy that is used as a medical tracer to track, for example, the flow of blood. It is used to treat millions of patients annually.

By bombarding deuterium gas as target with a deuteron beam from the Crocker cyclotron, Luis Alvarez[18] and his colleagues managed to identify a gas that had a long lifetime and chemically behaves like hydrogen gas. It turned out to be a hydrogen isotope with a nucleus containing one proton and two neutrons, which was named *tritium*. It played a major role in the development of nuclear weapons, but is important today for the development of fusion reactors, such as ITER,[19] and as a tracer to track the flow of ground water.

Irradiating uranium with neutrons, Edwin McMillan[20] and Philip Abelson used elaborate chemical analysis to identify *neptunium*, the first *transuranium element*. A little later, Glenn Seaborg and collaborators identified *plutonium*, another transuranium[21] element, that turned out to play a crucial role in the later development of atomic bombs.

Stimulated by the success of the cyclotrons, Lawrence proposed an even larger machine with a diameter of 184 inches. But halfway into its construction, the US were pulled into the Second World War.

Calutrons

Soon after Einstein's fateful letter[22] to US President Roosevelt in which he warned of the consequences of nuclear fission, earlier discovered by Hahn, Meitner, and Strassmann in Germany, the US launched the *Manhattan project* to develop an atomic bomb. At the time it was already known that the most abundant uranium isotope uranium-238 is less suitable than the much rarer isotope uranium-235. So, Lawrence decided to use the cyclotron to separate the two isotopes. Due to their different masses they travel on slightly displaced trajectories in the cyclotron. Collecting the fraction that arrives on the "light" side, the relative abundance of uranium-235 is slightly increased. Repeating the process multiple times enriches it to make it usable for atomic bombs. Early 1942, just after the US entered the war following the attack on Pearl Harbor, Lawrence had isolated a few tens of micrograms. This convinced the government to fund the construction of a battery of so-called *calutrons* that eventually produced the enriched uranium for the atomic bomb that destroyed Hiroshima.

Soon after these traumatic events the war ended and Lawrence could direct his attention to the completion of the large 184-inch cyclotron. Thanks to two great ideas that were hatched during the war, its beam quality was improved and its maximum energy greatly increased.

Great Idea: Weak Focusing

In early cyclotrons reaching the highest energies was the prime objective, while little attention was paid to the beam sizes. The beams grew significantly such that much of the beam was lost along the way. Luckily a way to confine the beams was discovered in the 1940s. Let's find out how that works.

The field of the cyclotron magnet forces the particles on a spiral, near-circular trajectory; let's call this the reference trajectory. For simplicity, let's assume it is circular and given by the black circle on the left-hand side in Fig. 4.5. Ideally we want all particles to follow this trajectory, even if they are perturbed slightly, for example, due to collisions with remaining gas molecules or small errors in the magnetic field. The dashed red line shows an example of such a perturbed trajectory. But the field is constant everywhere inside the magnet. So even the perturbed trajectory follows a circle with the same diameter as the reference and therefore it never strays far away. As an example, consider a particle that deviates from the reference at the bottom left of the graph. A quarter of a revolution later, near the top left, it crosses the reference. This behavior to bend far-away particle back to the reference is not very strong and therefore called *weak focusing*. Nevertheless it helps to contain the horizontal excursions of particles and thereby also the size of a beam as an ensemble of many particles.

All by itself, weak focusing only works in the horizontal plane, but we also need a restoring force in the vertical direction to keep the particles nicely centered between the magnet poles. This is accomplished by increasing the distance between upper and lower yoke of the magnet towards the outside, as shown on the right-hand side in Fig. 4.5. This causes the magnetic field lines to bulge outwards. From Crookes' experiments in Chap. 2 we know that the magnetic force points into the direction perpendicular to the field lines and that turns out to be back towards the mid-plane of the magnet. It is thus also focusing.

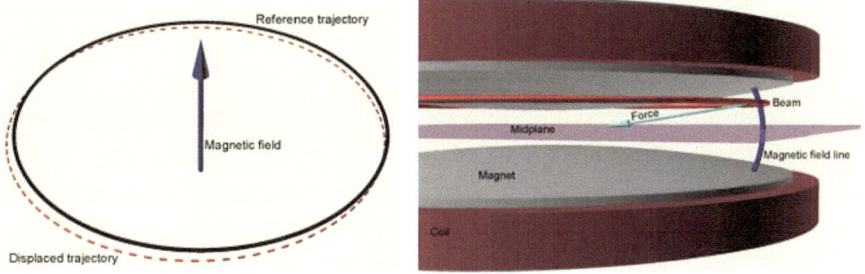

Fig. 4.5 Weak focusing

We cannot, however, increase the vertical focusing without limit by widening the distance between upper and lower yoke, because the field at larger radii becomes weaker. But we also need it strong to focus the beam horizontally. We therefore have to strike a balance between vertical and horizontal focusing, which was already weak to start with. As a consequence of this weak effect the particles in the beam are only ever so softly nudged towards their reference trajectory. This nudging prevents the beam from growing uncontrollably, but individual particles can still assume quite large deviations and this makes the beam size still quite large. As a consequence the vacuum chamber must be quite large and that forces the gap between the upper and lower magnet yoke to be big. And that makes the magnets very large, very expensive, and very hungry for electric power.

Originally, weak focusing was analyzed when building the first *betatrons* during the 1940s. Betatrons[23] are circular electron accelerators, whose rapidly increasing magnetic field causes an electric field that is used to accelerate the electrons. Donald Kerst[24] built the first working prototype and optimized its performance. He soon discovered that shaping the field as discussed in the previous paragraphs helps to keep the electrons close to the desired orbit.

Since there are restoring forces in both horizontal and vertical direction, the electrons perform oscillations around the reference trajectory, much like a spring providing a restoring force to a small mass that oscillates around its equilibrium point once it is perturbed. Practically all stable systems that are perturbed from their equilibrium position perform such oscillations, and accelerators are not different. If designed properly, particles in a beam perform stable oscillations around their reference trajectory. If these oscillations occur in the horizontal or vertical plane they are called *betatron oscillations*. After all, they were first observed in a betatron.

Weak focusing prevents the beam sizes from growing uncontrollably, such that it stays within the beam pipe, but all by itself it does not help to reach higher energies. That required another great idea.

Great Idea: Phase Focusing

The successful operation of cyclotrons with a fixed-frequency radio-frequency system crucially depends on all particles having the same revolution time, independent of their energy. The energy gained in each traversal from one dee to the other increases the speed of the particles such that it precisely balances the increased length of the trajectory. This works at energies below approximately 20 MeV for protons, at which point the effects of Einstein's special relativity become important. Increasing the energy does not only increase the speed of the particle but also its mass. This causes the particle to travel at a larger radius, which takes longer to traverse. The particle thus comes too late to the point where its energy is due to be increased.

Around 1945, both Vladimir Veksler[25] in the Soviet Union and Edwin McMillan in the US independently found that adjusting the radio-frequency system to the longer travel time actually captures and guides the particles along to higher energies. Even particles with slightly incorrect energies follow the acceleration process, because the radio-frequency system provides a restoring force, that gives rise to oscillations. They are commonly called *synchrotron oscillations*.[26]

Cyclotrons that adjust their radio-frequency system to compensate the effects of relativity are called *synchrocyclotrons*. Initially the radio-frequency is adjusted to capture a beam from the source in the center of the cyclotron. Then, as the energy of the particles increases, the frequency is slowly reduced such that part of the captured beam is guided towards higher energies. During this time no new beam is captured, only a small fraction makes it to high energies. Compared to a conventional cyclotron, the total number of accelerated particles is much lower, but much higher energies can be reached in synchrocyclotrons.

Both of these great ideas were used in Lawrence's 184-inch cyclotron in Berkeley, which became operational in 1946.

184-inch Cyclotron

The construction of the cyclotron already started in 1940 with the assembly of the magnet weighing 4000 tons. The fields were excited through copper coils, weighing another 340 tons. At the time such massive magnets could not be directly powered from the electrical grid. Instead a motor driven generator was used to excite the coils. After the US entered the war, the magnet was used in a pilot facility to separate uranium isotopes for atomic bombs. It served this purpose until the end of the war, whence the original construction was adapted to exploit the two great ideas. The pole faces of the magnet were shaped to provide both horizontal and vertical focusing for the beam and the radio-frequency system could change its frequency from 36 MHz down to 18 MHz, which made operation as a synchrocyclotron possible. The particles to accelerate are prepared by leaking gas—hydrogen, deuterium, or helium—into an electric arc which removes the electrons from the gas atoms. The charged particles then escape through a small hole into the region where the radio-frequency system accelerates them to higher energies. In this way the cyclotron delivered 64 pulses every second of protons with energies up to

730 MeV, deuterons with total energies up to 460 MeV, and alpha particles up to 910 MeV. First, targets were directly inserted into the cyclotron but later additional magnetic and electric deflectors were installed that allowed to extract the full beam into an external beam line, where it served multiple users.

By 1948, the 184-inch cyclotron was powerful enough to produce pions that Powell had earlier seen on photographic emulsions exposed to cosmic radiation (Chap. 3). Moreover, the high beam intensity in the cyclotron produced many more such tracks. The beams were produced by directing alpha particles accelerated to an energy of 300 MeV onto a carbon target. The negative constituents of the nuclear debris passed through part of the magnetic field of the cyclotron that bends them outward, opposite to the accelerated alphas. The debris was directed to a stack of photographic emulsions, which were exposed to the beam for about ten minutes. A particle's momentum could be determined from the curvature of a track. Folding in the thickness of the track made it possible to estimate the mass of the particle, which turned out to be that of the negatively charged pion, thus confirming Powell's discovery.

A year later even a neutral pion, labeled π^0, was found in the 184-inch cyclotron when R. Björklund smashed protons into different targets and observed the photons coming out. At an incident proton energy of about 290 MeV he found a threshold at which many pairs of 70 MeV gamma rays (which is just another name for very short-wavelength photons) appeared. This was interpreted as evidence for the appearance of a neutral particle with a mass corresponding to the energy of the two gammas, or 140 MeV, a value close to that of the charged pions.

More on Cyclotrons and More on Pions

Following the 184-inch cyclotron in Berkeley many synchrocyclotrons were constructed all over the world, even in Uppsala[27] in Sweden, the town where I live. But let us focus on the cyclotron in Chicago, first operational in 1951, that accelerated protons up to energies of 450 MeV. Impinging the accelerated protons onto an internal beryllium target produced pions that were extracted through holes in the shielding to the experimental area, where experiments with the pions were done. Negative pions deflect away from the direction of the proton beam and escape in the forward direction. Positively charged pions, on the other hand, were only deflected outwards provided they escape the target in the backwards direction, albeit at lower energy and intensity. Both types of pions were then guided to external hydrogen targets, where the interaction probability of pions with protons, the nuclei of hydrogen atoms, was determined by measuring their attenuation. The probabilities of negative and positive pions turned out to be significantly different. This was interpreted as an indication for the formation of a transient compound of proton and pion. Later this was confirmed to be the case, but the compound lived too short to be called a particle. Instead it was called a resonance and the particular one observed in Chicago was called Delta-resonance, the first of many more resonances to come.

The V-particles that spontaneously decayed into two charged tracks, as discussed near the end of Chap. 3, and puzzled the community remained out of reach for cyclotrons and even for the higher-energy synchrocyclotrons. Clearly, accelerators producing beams with much higher energies were needed.

Notes

1. Sir Ernest Rutherford at the anniversary meeting of the Royal Society, November 30, 1927. Proceedings of the Royal Society of London B102: 239–255.
2. John Cockroft (1897–1967) and Ernest Walton (1903–1995) received the Nobel Prize in Physics in 1951 "for their pioneer work on the transmutation of atomic nuclei by artificially accelerated atomic particles."
3. Apart from his explanation of the emission of alpha particles through quantum tunneling, George Gamow (1904–1968) contributed to the understanding of how elements, heavier than hydrogen and helium, are forged in stars.
4. The circuit consists of a sequence of capacitors and rectifying diodes. It was already conceived by Greinacher in 1914, but later re-discovered and extended by Cockroft and Walton for their accelerator.
5. Robert van de Graaff (1901–1967) invented the electro-static accelerator named after him in 1929. Later, he doubled the achievable particle energies in a so-called *tandem accelerator*. It first accelerates negatively charged ions and then rips off their electrons at high energy. This turns the negative into positive ions that are accelerated once again, thus at least (depending on the number of electrons ripped off) doubling their energy.
6. Gustaf Ising (1883–1960) was a Swedish physicist, best known for his invention of the linear accelerator.
7. Rolf Wideroe (1902–1996) not only built the first linear accelerator, but also conceived the first betatron (in this chapter) and colliding beam accelerators (Chap. 8). His life's story is told in: Pedro Waloschek, *The Infancy of Particle Accelerators: Life and Work of Rolf Wideroe*, Vieweg, Wiesbaden, 1994.
8. Ernest Lawrence (1901–1958) received the Nobel Prize in Physics 1939 "for the invention and development of the cyclotron and for results obtained with it, especially with regard to artificial radioactive elements." During the Second World War he was instrumental in enriching uranium-135 to make it usable for bombs. His lab in Berkeley became one of the hot-spots for the development of accelerators, and nuclear sciences.

9. Much of the early history of accelerators was shaped at Lawrence's lab in Berkeley, whose history can be found in: J. Heilbron, R. Seidel, *Lawrence and his Laboratory, Vol I,* University of California Press, Berkley, 1989. Available online at https://publishing.cdlib.org/ucpressebooks/view?docId =ft5s200764.

10. Stanley Livingston (1905–1986) built the first cyclotron for his Ph.D. He famously quipped that all he got was a Ph.D. whereas Lawrence received a Nobel prize for building the first cyclotron. Later he led the construction of the Cosmotron (Chap. 5). Together with Courant and Snyder, he invented the concept of alternating-gradient focusing (Chap. 6).

11. A H_2^+ molecule is composed of two protons and a single electron.

12. A deuteron is a nucleus composed of one proton and one neutron, as shown in Fig. 3.5. The chemical element with one electron orbiting a deuteron nucleus is called deuterium. It is chemically very similar to hydrogen.

13. When Ernest and John's mother was diagnosed with terminal cancer of the uterus in 1937, they brought her to Berkeley and organized her treatment with X-rays. She lived for another 15 years.

14. Martin Kamen (1913–2002) and Sam Ruben (1913–1943) were the first to synthesize the isotope carbon-14 in 1940.

15. Willard Libby received the Nobel Prize in Chemistry in 1960 "for his method to use carbon-14 for age determination in archaeology, geology, geophysics, and other branches of science." See his Nobel lecture for the background story. It is online available from https://www.nobelprize.org/prizes/chemistry/1960/libby/lecture/.

16. By ordering elements according to their chemical properties, Dmitri Mendeleev (1834–1907) created the first periodic table of the elements. Notably, he had to leave a few spots empty that were filled much later.

17. Emilio Segre (1905–1989) first identified the element technetium and, after moving to Berkeley, discovered the anti-proton in the Bevatron (Chap. 5). Jointly with his collaborator Owen Chamberlain (1920–2006) he received the Nobel Prize in Physics in 1959 "for their discovery of the antiproton."

18. Luis Alvarez (1911–1988) isolated tritium, invented an improved version (Chap. 5) of Wideroe's linear accelerator, improved the bubble chamber (Chap. 6), and, jointly with his son Walter, figured out that a large meteor impact on the Yucatan peninsula wiped out the dinosaurs 65 million years ago. He received the Nobel Prize in Physics in 1968 "for his decisive contributions to elementary particle physics, in particular the discovery of a large number of resonance states, made possible through his development of the technique of using hydrogen bubble chamber and data analysis." He tells his story himself in: Luis Alvarez, *Alvarez, adventures of a physicist,* Basic books, New York, 1987.

19. The aim of the *International Thermonuclear Experimental Reactor* (ITER) is the generation of energy by nuclear fusion, the process by which our sun generates energy. The facility is located in the South of France. For up-to-date information, see: Michel Claessens, *ITER: The Giant Fusion Reactor,* Copernicus Books Cham, 2023.

20. Edwin McMillan (1907–1991) invented the concept of phase focusing that allowed accelerators to reach much higher energies than before, as discussed later in this chapter. He also identified neptunium, the first element heavier than uranium and became director of the lab after Lawrence suddenly died. Jointly with Glenn Seaborg (1912–1999) he received the Nobel Prize in Chemistry in 1951 "for their discoveries in the chemistry of transuranium elements."

21. Neptunium and plutonium with element numbers 93 and 94 were the first elements heavier than uranium. Many more were to come later. Today the heaviest element, Oganesson, has number 118. Many accelerators played a prominent role in their discovery with main players located in Berkeley in the US, in Dubna—first in the Soviet Union then in Russia—and in Darmstadt in Germany. The early history of these discoveries is described in: Glenn Seaborg and Isadore Perlman, *The synthetic elements I,* Scientific American, April 1950, page 38; Glenn Seaborg and Albert Ghiorso, *The synthetic elements II,* Scientific American, April 1956, page 66; Glenn Seaborg and Arnold Fritsch, *The synthetic elements III,* Scientific American, April 1963, page 68; Glenn Seaborg and Justin Bloom, *The synthetic elements IV,* Scientific American, April 1969, page 56. For an up-to-date perspective see: Kit Chapman, *Superheavy,* Bloomsbury, London, 2019.

22. After having learned of Hahn's and Strassmann's discovery of nuclear fission, a group of European scientists (Leo Szilard, Edward Teller, Eugene Wigner) that had emigrated to the US prepared a letter warning of a German atomic bomb. Einstein, another emigree by the time, agreed to sign this letter and address it to US President Roosevelt, who initiated the *Manhattan project* to develop an American atomic bomb. The Germans did not succeed, but the Americans did and dropped bombs on Hiroshima and Nagasaki. For the full account of the history around the creation of the atomic bombs see: Richard Rhodes, *The Making of the Atomic Bomb,* Simon and Schuster, New York, 1986 and Robert Jungk, *Brighter than a Thousand Suns,* Harcourt Brace and Company, San Diego, 1958.

23. A. Wildhagen, *The Betatron,* Scientific American, May 1943, page 207.

24. In the process of building high-intensity X-ray sources, Donald Kerst (1911–1993) constructed the first operational betatron, an accelerator in which a rapidly increasing magnetic field causes (by induction) an electric field that is used to accelerate electrons. He carefully analyzed the shape of the magnetic field and realized that increasing the distance between upper and lower yoke of the magnet helps to keep the electrons in midplane of the magnet. For a description by the inventor himself, see: Donald Kerst, *Development of the Betatron and Applications of High Energy Betatron Radiations,* American Scientist, January 1947, page 56.

25. Vladimir Veksler (1907–1966) came up with the idea of phase focusing at about the same time as McMillan. In 1956 he became the founding director of the Joint Institute for Nuclear Research (JINR) in Dubna, where he supervised the construction of the *Synchrophasotron.*

26. This works because particles that have a little bit too high energy travel on the outside of the reference trajectory and have a longer way to go and therefore arrive a little too late at the gap between the dees. At that time the accelerating field is

already a little lower and the particle receives less energy. Little by little its energy approaches what it should be. Conversely, a particle with a slightly too little energy takes a shortcut and travels on the inside of the reference trajectory. It therefore arrives a little too early at the gap between the dees and receives slightly more energy. Again, little by little its energy approaches that of the reference particle.

27. The story of the Uppsala cyclotron is quite colorful. In 1926, The Svedberg received the Nobel prize in Chemistry "for his work on disperse systems." During the 1940s he convinced the clothing manufacturer Gustaf Werner that a cyclotron would be useful to explore new synthetic materials such a nylon. Werner agreed and provided funding. A colleague was promptly dispatched to Lawrence's lab in Berkeley where he arrived just after phase focusing was discovered. He promptly convinced Svedberg to build the cyclotron based on the new ideas. For a short while the cyclotron in Uppsala was the highest-energy accelerator in Europe and was actually used to analyze synthetic fabrics. After the report was handed over to Werner, he donated the cyclotron to the university, where it was used for nuclear physics and cancer therapy until 2016.

5

The Cosmotron Meets the Strangeness of Physics

After the Second World War, physicists were well reputed by the upper echelon of the newly formed *Atomic Energy Commission* (AEC), who assumed stewardship of all things nuclear in the US. After all, physicists had decisively contributed to the development of radar, a method to locate approaching aircraft by detecting radio waves that bounce off of aircraft. And, more importantly, they played an essential role in the development of the atomic bomb. In the process, they developed nuclear reactors, first to breed plutonium for atomic bombs and later to generate electricity. So, it is no surprise that during the post-war period the general political climate in the US favored the construction of the next generation of particle accelerators, both to maintain the technological advantage over the emerging new opponent, the Soviet Union, and to maintain a pool of well-trained scientists that proved to be so inventive in the past.

Since the existing synchrocyclotron already used enormous amounts of iron, enough to build large ships, a new idea was needed to increase the achievable energy of the particles significantly, at least by a factor 5, better 10, or even more.

Great Idea: The Synchrotron

And that great idea came from Mark Oliphant,[1] who in 1943 was overseeing the operation of gas centrifuges for uranium-isotope separation at Oak Ridge Laboratory. In parallel to his official duties he was thinking about how to increase the energy of accelerators. His idea was to confine the trajectory of the particles to a fixed radius and increase the magnetic field as the energy and

© The Author(s), under exclusive license to Springer Nature Switzerland AG 2024
V. Ziemann, *Beams*, Copernicus Books,
https://doi.org/10.1007/978-3-031-51852-2_5

49

the speed of the particles increases. As their speed increases, one turn in the accelerator takes less time, which makes it necessary to adjust the frequency of the acceleration system. Adjusting this radio-frequency system in synchronism with the magnetic field gave this type of accelerators its name: *synchrotron*. But by restricting the particles to a specific radius, it would be possible to leave out all the iron in the center of the accelerator, which would save an enormous amount of material. The iron would no longer be an impediment to reaching higher energies. But on the downside, it would be necessary to externally inject particles with some moderate initial energy into the magnet ring. This was deemed considerably more difficult than placing a particle source in the center of cyclotrons. Moreover, keeping the rising magnetic field and the radio-frequency system synchronized would require rather elaborate electronics. Oliphant's superiors received his idea rather coolly, especially because it was unclear, whether such a ring could operate in a stable way—phase-focusing was not yet invented at the time.

But by 1948 phase-focusing was well established and the AEC agreed to actually fund two large accelerators, both of them synchrotrons; one at Lawrence's laboratory in Berkeley on the west coast and the other at the recently established *Brookhaven National Laboratory* (BNL) on Long Island on the east coast.[2] The latter was named *Cosmotron*.

Cosmotron

Under the leadership of Lawrence's former doctoral student, Stan Livingston, the Cosmotron, with its circumference of 70 m, was taken into operation in 1953. It reached proton energies of up to 3300 MeV, almost ten times more than the cyclotrons and yet, it required only about half as much iron as the cyclotron in Berkeley. In previous accelerators the magnetic field along the beam's trajectory was the same. In the Cosmotron, on the other hand, some space was needed to accommodate the injection, to install radio-frequency components, and for the target to produce secondary particles for experiments. Therefore, as shown in Fig. 5.1, 3 m long straight and field-free sections were left between the four quadrants filled with magnets. This had not been done before and people worried about the stability of the beam until Ernest Courant and Hartland Snyder discovered an elegant way[3] to analyze the problem. Once they established that the stability of the beam, despite the field-free regions, should not be a problem, the Cosmotron was built and 3.5 MeV protons were injected from an external van de Graaff electro-static accelerator (Chap. 4). The magnetic field at injection was low, but was then ramped up to its peak value within one second. Like in cyclotrons before and to minimize collisions with

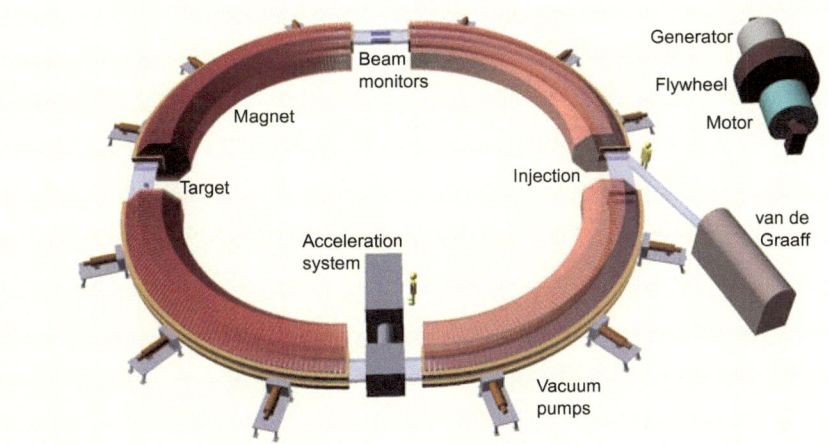

Fig. 5.1 Cosmotron

gas molecules, the beam traveled in a vacuum chamber sandwiched between the magnet yokes. Beam monitors carefully tracked the beam's position and their signal was used to fine-tune the magnets. Weak focusing confined the beam, but as the name implies, only weakly such that the beam size was still rather large. Moreover, finite tolerances when controlling the beam position caused additional excursions of the beam inside the vacuum chamber, which therefore had to be rather large. It had a height of 22 cm and was 90 cm wide. Consequently, the magnets had to be very large and required an enormous amount of power to excite them to maximum field within one second. This power was provided by a generator powered from a motor-driven 43-ton flywheel rotating at 900 revolutions per minute. Once the field reached its maximum value the beam with an intensity of about 10^{10} protons was directed onto a target for experiments. After a short time at the maximum energy the magnet was ramped down and about 75 % of the energy from the magnets was recovered by speeding up the flywheel. The remainder was compensated by the continuously operated motor. Such a cycle was typically repeated every five seconds.[4]

Since the speed of the protons varies about ten-fold between injection and its maximum value, the frequency of the RF system had to increase by the same factor. Moreover, it had to be synchronized to the rise of the magnetic field with high precision. This was achieved by measuring the field with a small coil and converting the induced voltage with a vacuum-tube-based electronic circuit to a control signal for the frequency generator. All this was accomplished without transistors or digital computers—a remarkable feat!

Finally, an accelerator capable of producing and exploring the ominous V-particles was available. First experiments used a beryllium target inserted from the inside of the ring. Turning off the radio-frequency system after the peak energy was reached caused the circulating protons to slowly lose energy such that they spiral inwards and strike the target. Neutrons, emitted from the collisions and unaffected by the magnet, left the target in the forward direction, where they produced V-particles that were observed in cloud chambers. At last more than just a few images of these reactions became available such that a study of their characteristic behavior could start. In another experiment, negative pions, emitted from the protons hitting a carbon target, were extracted and directed through holes in the shielding walls towards a cloud chamber. This produced even more V-particles. After a few years the accelerator crew even learned how to extract the full-energy proton beam and direct it onto external targets where it produced a wide variety of secondary particles, including charged versions of the ominous V-particles. A large magnet was then used to select them according to their energy and charge. In the process, many more particles even heavier than protons were found in cloud chambers and photographic emulsions.

Associate Production and Strangeness

From these first measurements of accelerator-produced V-particles it appeared that they always show up in pairs. Many cloud-chamber photos showed two of the characteristic V-shaped events where invisible particles decay into two charged, and therefore visible, daughter products, as shown in Fig. 5.2. Abraham Pais gave this dual appearance of V-particles the name *associate production* and Murray Gell-Mann[5] conjectured that an inherent property of the V-particles, he called it *strangeness*, is preserved in the strong nuclear force that is responsible for their creation. The protons from the beam, however, would not have this property and neither would the nucleons in the target material have it. Therefore two V-particles, one with a positive unit of strangeness and another one with a negative unit of strangeness, must be created. Pais' associate production and Gell-Mann's strangeness, though introduced ad-hoc, turned out to be tremendously useful. These concepts provided both a classification scheme and a conservation law.[6]

Much of particle physics is based on inventing classification schemes and finding quantities that are conserved in various reaction. One obvious example is the conservation of energy which all reactions have to obey. Another, albeit somewhat less intuitive one is the conservation of momentum; it helps us to understand how billiard balls move after a collision. Elementary particles have

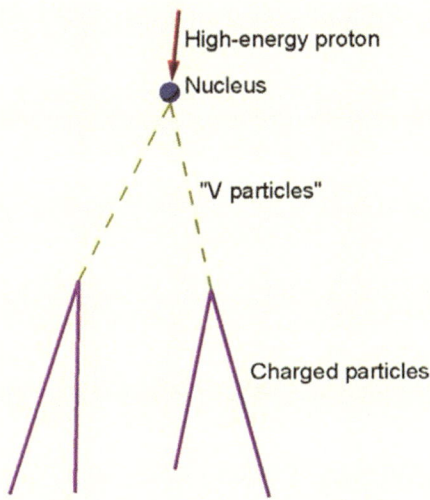

Fig. 5.2   Associate production

to obey these same rules; they are commonly referred to as *conservation laws.* Gell-Mann and others tried to find additional laws that govern the interactions of elementary particles. But to do so, they first need to order—classify—the particles into groups that behave similarly. That's the origin of putting electrons and muons into one group—leptons—and pions, protons, and neutrons into another one—hadrons. Once the particles are grouped, Gell-Mann and his colleagues tried to invent new quantities that do not change in reactions; strangeness is such a new conserved quantity that affects all particles in one group and is consistent with experimental observations such as the one shown in Fig. 5.2. Thus they arrive at a conservation law for strangeness in the reaction that created the V-particles. But strangeness is not conserved in the decay of the V-particle, because the charged decay products turned out to be particles without strangeness that were known before. Therefore some other process had to be responsible for the decay and this turned out to be the weak interaction. Much of what Gell-Mann and his colleagues did was trying out different classification schemes and finding a framework of "laws" that is consistent with all experimental observations. It's a lot like laying a large puzzle without initially understanding what the picture is. But Gell-Mann and colleagues were tremendously successful and figured out how the pieces fit together.

It even gave some indication about the lifetime of V-particles. If the strong nuclear force were responsible for their decay, the tracks they leave would have been too short to be visible on cloud chamber images. Since we can actually see them, some other fundamental force must be responsible for their decay and the weak force, normally associated with radioactivity, appeared to be a hot candidate. We will pick up this thread in a short while, because the Cosmotron was not the only accelerator involved in solving the strangeness puzzle. Soon after the Cosmotron, the second large synchrotron funded by the AEC, the *Bevatron* in Berkeley, became operational.

Bevatron

The Bevatron,[7] with its circumference of 120 m and maximum energy of 6200 MeV, was almost twice as large as the Cosmotron. After a delay,[8] construction of the Bevatron got under way and the machine saw first beam in 1954.

Protons and also other particles were generated in an external ion source and accelerated by a Cockroft-Walton accelerator to 0.45 MeV, where a linear accelerator, designed by Luis Alvarez and shown in Fig. 5.3, took over and boosted their energy to 10 MeV. This linac is based on Ising's and Wideroe's drift tubes, but encloses the sequence of tubes inside a larger vacuum tank that is tuned to resonate at a frequency of 200 MHz. An antenna feeds microwaves with that frequency into the tank such that standing waves are excited inside. Provided the drift tubes inside the tank are at the right place, the particles, shown in red, can "hide" inside while the electric field changes polarity and accelerates the particles again between adjacent drift tubes. Many modern-day linear accelerators for protons and heavy ions, after almost 80 years, are still based on Alvarez's idea.

Like the Cosmotron, the Bevatron contained four magnet-filled quadrants with intermediate field-free sections. One of them was used to inject 10 MeV protons ten times per minute, just before the magnets were excited by a flywheel-driven generator. Within 2 s, the magnets reached their maximum field and protons their maximum energy, ready to strike a target. Initially

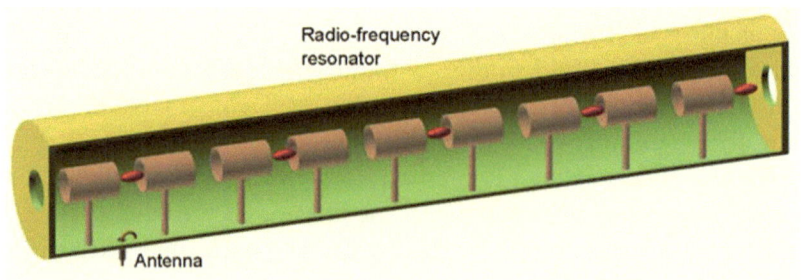

Fig. 5.3 Alvarez's drift-tube linear accelerator

the design of the Bevatron was very conservative with a 1.2 m gap between upper and lower magnet yoke, leaving plenty space for the weakly focused proton beams, but after the successful operation of the Cosmotron, the gap was reduced to 30 cm. This permitted the machine to actually reach the desired 6200 MeV peak energy for protons that was needed to produce *antiprotons*.

Antiprotons

So what's the fuzz about antiprotons? Two decades earlier Dirac had predicted the existence of positrons, the anti-particles of electrons, and a little later Anderson actually found them (Chap. 3). So, it is not surprising that people started to wonder whether also other particles, especially those heavier than electrons, have corresponding anti-partners. Since creating heavy particles required much higher energies than was available in cyclotrons, this could only be done with a new class of accelerators. The Bevatron was the first one capable of producing them.

The high-energy protons in the Bevatron were intercepted by a copper target, shown on the top-right in Fig. 5.4, that created a spray of nuclear debris from which the antiprotons were filtered out.[9] Some focusing magnets are needed to prevent the particles from going astray while the deflecting magnets determine their momentum. Since momentum is the product of a particle's mass and velocity, both lower-speed heavy antiprotons and the lighter negative pions with higher speed follow the same path. In order to resolve this ambiguity Emilio Segre and Owen Chamberlain[10] used two scintillators[11] that produce short pulses to determine the speed of particles, such that they could discriminate pions from antiprotons. Additional watchdog signals ensured that at the end of the beam line only antiprotons were counted with electronic circuitry. Optionally directing these isolated antiprotons onto photographic emulsions caused proton-antiproton annihilations. They show up as a distinct

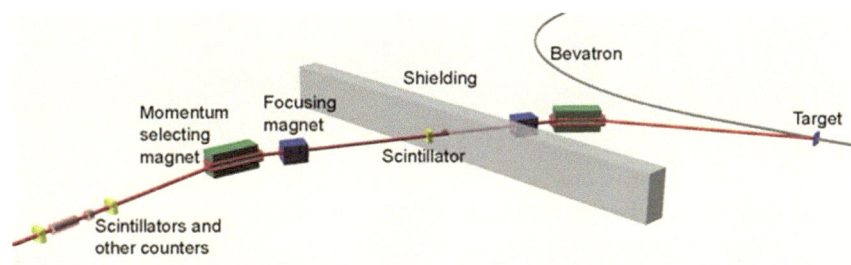

Fig. 5.4   Antiproton discovery

"star" on the image from which charged particles emanate. The energies of these particles added up[12] to that of the antiproton and the stationary proton in the emulsion, which added weight to the discovery of the antiproton.

The antiproton was probably the biggest fruit that the Bevatron picked, but not the only one. It joined the race to figure out the puzzle of these strange V-particles. In particular, placing a target in the field-free region between quadrants allowed all particles, irrespective of their charge, to escape. Soon charged V-particles could be isolated in external beam lines, rather similar to the one used to detect antiprotons. Passing them to detectors then allowed the experimenters to investigate the nuclear reactions they caused. Moreover, directing the extracted full-energy proton beam onto targets with subsequent momentum analysis produced even more particles whose reactions were analyzed in a recently invented new detector—the *bubble chamber.*

Bubble Chamber

In bubble chambers[13] a liquid, for example the anesthetic ether, is heated to its boiling point before the volume is enlarged by pulling on the piston shown in Fig. 5.5. In this condition, part of the liquid is inclined to become a gas and forms bubbles, but can only do so if an inhomogeneity is available to trigger the bubble formation. Charged particles, passing through the liquid, provide such a trigger and cause bubbles to appear along their tracks. These tracks are very well-defined and their photographic images are suited to determine properties of the particles.

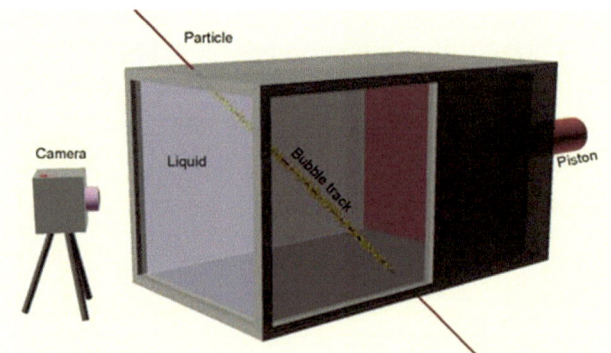

Fig. 5.5 Bubble chamber (track not to scale)

A distinct advantage of bubble chambers is the larger density of liquids compared to the gases in cloud chambers. Bubble chambers therefore also act as targets in which many more nuclear reactions occur. This remarkable device, using ether as a liquid, was invented by Donald Glaser.[14] Right after he presented his idea at a conference, Luis Alvarez suggested to replace ether by hydrogen, cooled to very low temperatures where it turns into a liquid. This has the additional advantage that the target material only contains the protons of the hydrogen nuclei, which gives rise to very well-defined reactions.

Finally, triggering the expansion of the liquid at the moment when the charged particles from the accelerator arrive, ensures that most photos actually contain nuclear reactions. As a result, bubble chambers were the instruments of choice for much of the 1960s and into the 1980s. The first one was installed at the Bevatron where it recorded large numbers of V-particles, both the neutral and therefore invisible kind that decayed into two charged particles, but also charged versions of either polarity. Jointly these particles were later named *Kaons*.

Kaons and the Tau-Theta Puzzle

With the proliferation of particles from the Bevatron and the Cosmotron, systematic studies of these particles became possible. They showed that the three types of Kaons, the neutral K^0, the positively charged K^+, and the negatively charged K^- all have about the same mass, around $500\,\text{MeV}/c^2$, roughly halfway between the mass of electrons and that of protons. They are therefore also classified as mesons.

The positively charged K^+ meson puzzled the community for a long time, because sometimes it decays into two pions ($\pi^+ + \pi^0$) and sometimes into three pions ($\pi^+ + \pi^+ + \pi^-$). The charges nicely balance, but the strangeness does not; the kaon carries strangeness and the pions do not. Moreover, at the time it was known that the pions carry a somewhat abstract intrinsic property called *parity*, which turned out to be another very useful classification scheme. It is negative for all pions, both neutral and charged. The math used to work with parity dictates that the parity values of two particles must be multiplied. The final state with two pions therefore has positive parity and the final state with three pions has negative parity. For a long time people thought that two different particles, called tau and theta, were responsible for the two- and three-pion decay, respectively. Yet, their masses and their lifetimes were the same when measured with increasing precision such that they must be the same particle that was now called K^+. That this particle had two decay modes with opposite parity was puzzling, which was called the tau-theta puzzle.

Chen-Ning Yang and Tsung-Dao Lee[15] suspected that the weak nuclear force, which is responsible for the decay of kaons, might not care about conserving the parity.

Parity

So, let us take a closer look at parity. It is a fundamental property, also called a "quantum number," that most fundamental particles possess; spin and strangeness, for example, are other such properties. Parity, however, is related to the mirror symmetry of a process. In a collision between a red and a white billiard ball, the red one can bounce either to the right or to the left. One case is just the mirror image of the other case. There is no fundamental law preventing either one. If we now could find a process that would make the red ball always bounce to the right, we could distinguish a process from its mirror image. In our experience and that of all physicists up to the mid-1950s, every process, even a subatomic one, had the property that the process or its mirror image was equally likely. This is referred to as "parity is conserved." If, on the other hand, an experiment shows a distinct preference for one outcome, compared to its mirror image, we say that "parity is violated." Around the second half of the 1950s evidence was mounting that the latter was the case.

This evidence came from experiments that exploited the spin of particles. Since spin has to do with the sense of rotation that we can visualize by considering metallic screws to connect pieces of wood. Technically, they conserve parity, because both clockwise and counterclockwise turning screws are possible. By convention, however, practically all screws move into the wood when turning them clockwise. But this is only a convention! No fundamental law prevents screws with the opposite sense. They are just not used very often. In the subatomic world, on the other hand, there is some law against one type of screw.

It turns out that the fundamental particles corresponding to the screws are the neutrinos that showed up in Fermi's analysis of the radioactive decay of neutrons. They practically move with the speed of light and they have an intrinsic sense of turn, their spin. This makes it possible to distinguish between neutrinos having their spin pointing in the direction of motion and those with spin pointing in the opposite direction. What is surprising is that only the latter, left-handed neutrinos take part in the weak interaction. What made Lee's and Yang's analysis so special is that they proposed experiments to find out whether a particular nuclear reaction or its mirror image occurs in reality.

They proposed[16] to align a radioactive isotope of cobalt, cobalt-60, in a very strong magnetic field and observe the direction of the emitted electrons. The

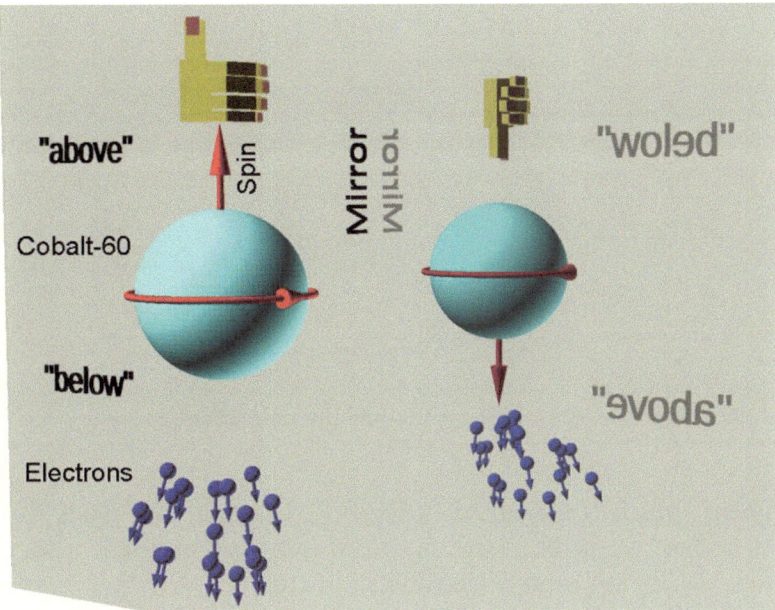

Fig. 5.6 Parity violation

spin of all cobalt atoms then aligns along the magnetic field lines and defines a "sense of rotation" for the cobalt atoms, which is indicated by the arrow near the waist line of the atoms in Fig. 5.6. Using the fingers of our right hand to point in the same direction as the arrow, the thumb defines a specific direction that we call "above." The opposite direction we then call "below." Detecting more electrons in one direction than the opposite direction would allow us to distinguish an experiment from its mirror image and parity would not be conserved.

Actually doing the experiment turned out to be fiendishly difficult because the cobalt atoms had to be cooled down to within a few parts in a thousandth of a degree of absolute zero temperature. But Chien-Shiung Wu[17] succeeded and found that electrons are only emitted into the direction opposite of the arrow that identifies the direction of cobalt atoms' direction of spin; that's "below." In the mirror world, on the other hand, the electrons, the sense of rotation is reversed and the spin therefore points downward. In this mirror world, the electrons would be emitted into the same direction as the spin, but that is never observed in our universe. This proved that parity is not conserved in weak interactions, such as radioactive beta decay.

Immediately after Wu's findings became known, a group around Leon Lederman[18] at Columbia University realized that they could do a complementary

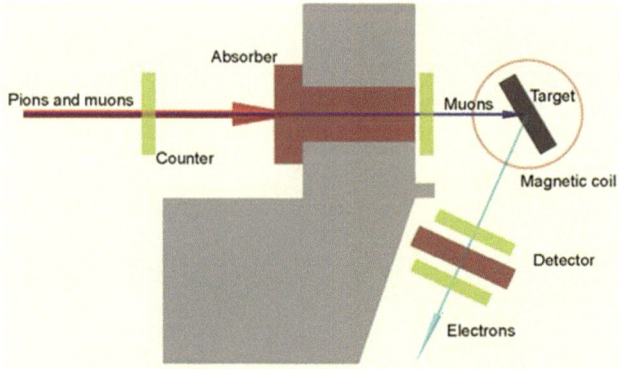

Fig. 5.7 Cyclotron-based experiment to show the violation of parity

experiment, which is illustrated in Fig. 5.7. They directed pions from their cyclotron towards a carbon target, exposed to a weak magnetic field. Most of the pions decay into muons within a short distance after their creation and most of the remaining pions were absorbed by a 20 cm long absorber. Only the muons made it through to the carbon target where they were stopped and given some time to decay into electrons and neutrinos. The arrival of particles was recorded by a number of counters. Since the muon has a magnetic moment, the magnet could systematically change the orientation of the incident muon. The electrons were detected and found to be emitted into a preferred direction, which indicated that the spin of the originating muons had a preferred orientation, again confirming the violation of parity in weak interactions. This accelerator-based confirmation of Wu's finding was done within days of Wu's experiment and appeared with her article back-to-back in the same journal.[19]

Parity was the first fundamental conservation law, held dear by physicists, that turned out to be violated in weak interactions. More was to come later. But first we need to explore particles with strangeness some more.

Resonances

The ability to produce external beams of pions, and later also of kaons, together with the larger number of useful images of events from bubble chambers, made it possible to produce many new particles and, on top of that, systematically investigate their properties.

Directing kaons from the Cosmotron or the Bevatron into bubble chambers produced a wealth of particles heavier than protons that also decayed very slowly. They seemed to inherit strangeness from their parental kaons and were called *hyperons.* Three new particles with different electric charges were discovered, all of them having the same mass: the neutral lambda Λ^0, three charge states of the sigma Σ^+, Σ^-, Σ^0, and two charge states of the xi, labeled Ξ^0, Ξ^-. Carefully keeping track of the strangeness of both parent particles and offspring made it possible to assign values of strangeness and other quantities, such as spin to these particles. For example, the Ξ^- was found to carry two negative units of strangeness and has the same spin as an electron. Most of these particles, at least the charged ones, live long enough to leave visible tracks in images.

But soon other particles, called *resonances,* were discovered.[20] Their lifetime is too short to leave tracks, but their existence was inferred from the observation of their decay products. Consider a particle interaction where one incoming particle produces three decay products, labeled A, B, and C. We then determine the energy of C. If the disintegration of the incoming particle happens instantaneously, the available energy can be distributed in many different ways and the energy of C can assume a wide range of values. If, on the other hand, A and B form a compound (AB) from which C recoils, the energy of C will always be the same. By observing the distribution of energies of decay products we can therefore infer the existence of a very short-lived compound from which the third decay product recoils. This short-lived compound is the resonance.

The first resonance found was the Delta-resonance (see end of Chap. 4) that, due to the limited energy, could be only partially resolved in the Chicago cyclotron. Closer inspection showed that this is a compound of a proton or neutron and a pion and has a larger mass than a proton. Moreover, four charge states of the Delta were identified: Δ^-, Δ^0, Δ^+, and Δ^{++}. Collectively, all these new particles and resonances with masses comparable and above that of protons are called *baryons*[21] in order to distinguish them from the lighter mesons. And more new particles showed up when painstakingly analyzing many thousands of bubble chamber images. In the Bevatron three short-lived mesons were found; the eta η and the omega ω which are neutral, and three charge states of the rho: ρ^+, ρ^-, ρ^0. As it turned out, these were just the first batch of new particles. Many more came later.

This plethora of particles just screams for a scheme to restore order in this "zoo of particles," as it was called. New and more powerful accelerators brought progress.

Notes

1. Mark Oliphant (1901–2000) played an important role in the development of radar and conceived the synchrotron. He later built a 1 GeV synchrotron in Birmingham, that started operation in 1953.

2. The Synchrophasotron in Dubna, reaching proton energies of up to 10 GeV in 1957, was the third synchrotron conceptually similar to the Cosmotron and the Bevatron.

3. Ernest Courant (1920–2020) and Hartland Snyder (1913–1962) developed a mathematical formalism to describe the motion of particles in accelerators that was used to design practically all of them ever since. One of the early applications of this formalism was to show that the gaps between magnets do not preclude stable operation and later that alternating-gradient focusing (Chap. 6) results in a much improved performance of accelerators. Their new method was based on representing all magnets and the space between magnets by mathematical quantities, called matrices. These matrices have the remarkable property that a sequence of elements is described by a matrix that is calculated from multiplying the matrices for each element. Furthermore, they could derive mathematical conditions for the matrix describing one full turn in the Cosmotron that guarantees stable motion of particles. They published their work in a highly influential report: E. Courant, H. Snyder, *Theory of the Alternating-Gradient Synchrotron,* Annals of Physics 3 (1958) 1.

4. The complete September issue of the scientific journal *Review of Scientific Instruments* from 1953 is dedicated to the Cosmotron and all its components that use the technology of the time with all electronics based on vacuum tubes.

5. Apart from introducing the concept of strangeness, Murray Gell-Mann (1929–2019) came up with the "eightfold way" to classify the "particle zoo" and the hypothesis of quarks (Chap. 6). After he received the Nobel Prize in Physics in 1969 "for his contributions and discoveries concerning the classification of elementary particles and their interactions," he introduced the concept of "color charge" in the emerging quantum chromodynamics (Chap. 7). His biography is told in: George Johnson, *Strange Beauty: Murray Gell-Mann and the Revolution in Twentieth-Century Physics,* Knopf, New York, 1999.

6. A vivid account of the developments at the time can be found in: Murray Gell-Mann and E Rosenbaum, *Elementary Particles,* Scientific American, July 1957, page 72.

7. Lloyd Smith, *The Bevatron,* Scientific American, February 1951, page 20.

8. The delay was caused by a secret and classified military project, called MTA, a large linear accelerator to create fissionable material. All key personnel disappeared from Berkeley for almost two years, before large quantities of Uranium were discovered in the US and made the MTA obsolete.

9. A first-hand account of the discovery of the antiproton is given in: Emilio Segre and Clyde Wiegand, *The Antiproton,* Scientific American, June 1956, page 37.

10. Emilio Segre (1905–1989) and Owen Chamberlain (1920–2006) received the Nobel Prize in Physics in 1959 "for their discovery of the antiproton."
11. A scintillator emits light as a consequence of a passing particle. The light is then converted by an electro-optical circuit to a signal that is handled by electronic circuitry.
12. Actually the energies and momenta of all particles are taken into account to calculate what is called the antiproton's "invariant mass", which corresponds to its mass at rest.
13. The bubble chamber is described by its inventor: Donald Glaser, *The Bubble Chamber*, Scientific American, February 1955, page 46.
14. Donald Glaser (1926–2013) received the Nobel Prize in Physics in 1960 "for the invention of the bubble chamber."
15. Chen-Ning Yang (b. 1922) and Tsung-Dao Lee (b. 1926) received the Nobel Prize in Physics in 1957 "for their penetrating investigation of the so-called parity laws which has led to important discoveries regarding the elementary particles." Jointly with Robert Mills did Yang formulate the so-called *Yang-Mills theories* (Chap. 6) which form the mathematical base for the standard model of particle physics.
16. For a more detailed and readable overview see: Philip Morrison, *Overthrow of Parity*, Scientific American, April 1957, page 45.
17. Chien-Shiung Wu (1912–1997) famously performed the experiment that proved the violation of parity for the first time.
18. After confirming the results of Wu's experiment on the violation of parity, Leon Lederman (1922–2018) proved that the muon neutrino is different from the electron neutrino (Chap. 6). He found the bottom-quark (Chap. 9), and became director of Fermilab (also Chap. 9). He received the Nobel Prize in Physics in 1988 "for the neutrino beam method and the demonstration of the doublet structure of the leptons through the discovery of the muon neutrino."
19. The two articles about the violation of parity appeared in *The Physical Review* 105 (1957) between page 1413 and 1417.
20. For a more detailed and readable overview written at the time see: R. Hill, *Resonance Particles*, Scientific American, January 1963, page 38.
21. Baryons are part of yet another classification scheme for particles. Their name is derived from the Greek word "baryos" for "heavy." Jointly, baryons and mesons make up the hadrons. In the next chapter we will see that this classification is very successful: baryons are made of three quarks and mesons of one quark and one antiquark. Stay tuned for the next chapter.

6

CERN and the Taming of the Zoo

In the post-war years of the late 1940s, a number of European scientists hatched the idea of a nuclear research facility, jointly funded and operated by participating member states.[1] By 1952 the discussion had matured and, sponsored by the *United Nations Educational, Scientific, and Cultural Organization* (UNESCO), eleven European countries signed a convention to establish a provisional organization of the *Conseil Européen pour la Recherche Nucléaire* (CERN), the European Council for Nuclear Research. Following an interim period, CERN was formally established in 1954.

Right from the start it was clear that the new organization should become the home of two top-notch particle accelerators: a 600 MeV synchrocyclotron, evolved from the 184-inch machine in Berkeley, and a proton synchrotron, similar to those in Brookhaven and Berkeley, but for higher energies. As experience with synchrotrons was scarce in Europe at the time, a delegation was sent to Brookhaven to participate in the dedication ceremony of the Cosmotron, mainly to discuss with and learn from American colleagues.[2]

The accelerator physicists in Brookhaven welcomed their European colleagues with open arms. Apart from just describing how they built the Cosmotron, they even mulled over ways to improved its design. And, oh boy, what a great idea they came up with. Stan Livingston, Ernest Courant, and Hartley Snyder figured out a way to reach much higher energies with much smaller magnets. They had invented *alternating-gradient focusing*.

© The Author(s), under exclusive license to Springer Nature Switzerland AG 2024
V. Ziemann, *Beams*, Copernicus Books,
https://doi.org/10.1007/978-3-031-51852-2_6

Great Idea: Alternating-Gradient Focusing

A problem that bothered Livingston was that all the magnets in the Cosmotron were open towards the outside of the ring. While this made extracting the negatively charged particles towards the outside of the ring easy, extracting positively charged particles towards the inside of the ring was prevented by the iron of the magnets. He therefore pondered whether some of the magnets could be installed backwards, with the open side facing towards the inside of the ring. Figure 6.1 illustrates his idea. Of course the opening of the magnets is responsible for focusing the beam (Chap. 4) and putting it into the ring backwards would mess this up. Courant and Snyder had earlier developed an elegant way to analyze the stability of trajectories and they almost immediately found that reversing some of the magnets would actually increase the stability of orbits. Moreover, increasing the widening of the magnet gap actually makes focusing even stronger.[3] The obvious benefit of this new, and stronger, focusing scheme are smaller beam sizes, potentially much smaller beam sizes, which then require much smaller magnets. And that opens up the way to higher energies: larger rings with many small magnets provide higher energies with less iron![4]

The new focusing scheme also brought new difficulties. First, the magnets had to be manufactured and aligned with much tighter tolerances than before. Often they had to be positioned with a precision given by the width of a hair in order to prevent the beam from becoming unstable and hitting the beam pipe. The second problem, called *transition energy crossing*, comes from a small energy dependence of the time to complete one revolution in the ring. If the energy of a proton is slightly higher than the reference, its speed is also higher and it arrives early. But once the speed of protons approaches the speed of light, Einstein's relativity dictates that the protons get heavier. Therefore they lie on the outside of the ring, have a greater distance to travel, and arrive later. At the transition energy, these two effects cancel each other and the protons

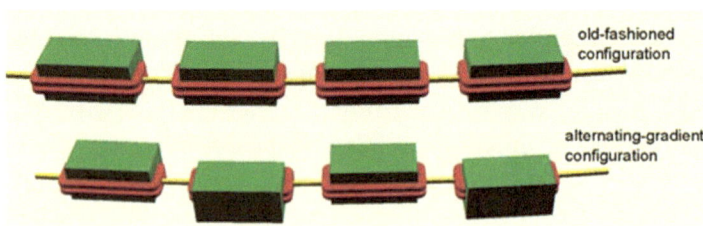

old-fashioned
configuration

alternating-gradient
configuration

Fig. 6.1 Old-fashioned and alternation-gradient configuration

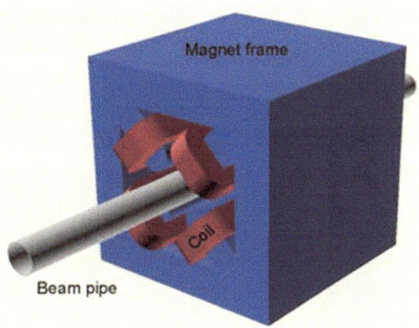

Fig. 6.2   Quadrupole

do not know what to do, so to say. Both difficulties were eventually solved by the ingenuity of the accelerator builders.

On the other hand, alternating-gradient focusing, or *strong focusing,* as it also became known, proved useful also in other contexts. Almost immediately John Blewett[5] and his colleagues in Brookhaven realized that they could also drastically extrapolate the increased focusing in a bending magnet to a point where a magnet only focused but no longer deflected the beam. Such magnets, a sketch is shown in Fig. 6.2, became known as *quadrupoles,* because they have four magnet poles instead of the two—upper and lower—poles of a bending magnet. Quadrupoles can focus a beam even in straight lines, such as in linear accelerators, where they have been used ever since.

Despite the difficulties, the prospect of strong focusing was too good to let it pass and plans to build a new generation of synchrotrons[6] with ten times higher energies immediately got under way, both in Brookhaven and at CERN.

CERN Proton Synchrotron

Coming back from Brookhaven, the CERN scientists immediately got to work and re-designed their synchrotron to reach 25 GeV, and later even 28 GeV. The ring were to have a circumference of 628 m, but with a vacuum chamber only 7 cm high, the iron of the magnets only weighed 3700 tons. This was roughly the weight of the Bevatron magnets, despite a six times larger circumference and reaching four times higher energies. Figure 6.3 gives an impression of the size of the accelerator and the large buildings to house the experiments. Like its predecessors the CERN proton synchrotron, or the PS for short, used a motor-flywheel-generator combo to power the magnets in 3 s long cycles. In each cycle the protons were generated in an ion source and pre-accelerated

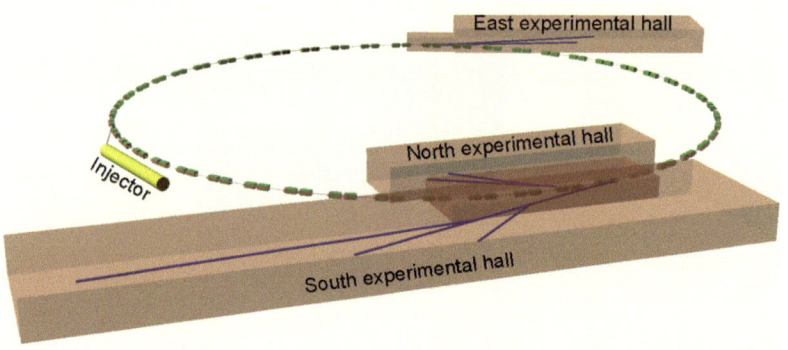

Fig. 6.3 CERN proton synchrotron

in a Cockroft–Walton to 0.5 MeV. At that energy an Alvarez-type drift-tube linac, aided by the novel quadrupole-based focusing, accelerated the protons to 50 MeV before they were injected into the PS.

Thanks to carefully aligning the PS during construction and placing the entire accelerator on a circular concrete slab with a 2 m × 2 m cross section, instabilities during the acceleration cycle could be avoided. Within a short time energies up to few GeV were reached but just as the transition energy was about to be crossed, the beam disappeared. Luckily Wolfgang Schnell quickly improvised an electronic control box in a, by now famous, Nestle-coffee can. It rapidly switched a parameter of the radio-frequency system, just as transition energy was crossed; all of a sudden[7] the beam made it up to 24 GeV. Some tuning then made it possible to reach the design performance of the PS early in 1960.

Initially, internal targets were used to create secondary particles which were directed to experimental areas with bubble chambers and other detectors. But exposing internal targets to high-energy beams created also large amounts of radioactivity that destroyed delicate equipment. Therefore new methods to extract the beam were implemented. Extracting the entire beam in one shot was one option, slowly peeling off a little beam at a time distributed over a few seconds was the second option. The extracted particles were routed to many experimental stations in several experimental halls. They passed through beam lines with combinations of magnets that filter out pions, kaons, or antiprotons at much higher energies than previously possible. The selected particle type was then directed to bubble chambers or photographic plates.

Shortly after the PS started operating and was put to use, also its twin, the alternating-gradient synchrotron (AGS) in Brookhaven, was commissioned.

AGS in Brookhaven

Remarkably, the application papers for funding the AGS were only five pages long and were approved after four months.[8] Today, projects of this scale require thousands of pages and are only approved after years of negotiations. Well, the crew in Brookhaven was lucky and could start to test the new focusing scheme and transition crossing on a small electron ring that used electric instead of magnetic guide fields. When convinced that they could handle these effects, construction of the AGS started. With a circumference of 807 m it is slightly larger than the PS and more than ten times larger than its predecessor, the Cosmotron, but with 33 GeV it also reached ten times higher energies. At the same time it required only about 4000 tons of iron, because the beams are much smaller and fitted into a much smaller vacuum chamber with a height of 8 cm.

An operation cycle of the AGS begins with creating protons in an ion source by burning hydrogen gas in a plasma. They are then accelerated to 0.75 MeV in a Cockroft–Walton, before an Alvarez-type drift-tube accelerator brought their energy up to 50 MeV, at which point they were injected into the AGS. Every five seconds the magnetic field in the AGS was low and suitable to accept the protons. Within a little over a second a motor-flywheel-generator combo generated enough electric power to excite the magnets to their peak field. Simultaneously, the radio-frequency system synchronously increased the energy of the protons such that the radius of their trajectory remains unchanged. After having reached their maximum energy, the protons either strike an internal target or are extracted into external beam lines, where they serve multiple experiments. Once the beam is used up, the energy stored in the magnets is redirected to the generator, which now acts as a motor and speeds up the flywheel, ready to power the magnets on the next cycle.

In the summer of 1960 the AGS was ready for experiments and three of the early ones struck gold.

One or Two Types of Neutrinos?

Following the discovery that parity was not conserved in the weak interaction, the latter became a hot topic. Especially the role of neutrinos was at the center of attention. The existence of neutrinos had been hypothesized by Pauli and Fermi earlier (Chap. 3). But finding them is extremely difficult, because they rarely participate in nuclear reactions. One needs enormously large numbers of neutrinos and very clever detection electronics to observe a few rare events. Fortunately, by the mid-1950s, nuclear power plants were producing copious numbers of neutrinos. Frederik Reines[9] and Clyde Cowan took on the challenge and built a fiendishly clever detector capable of observing a few neutrino-induced reactions, thus confirming their existence.

Coming from a nuclear power plant, it was known that Reines' and Cowan's neutrinos are born together with an electron. Now the question is whether electrons and their heavier relatives, the muons, share the same neutrino or whether each has its own? One way to find out is to produce large numbers of muon-neutrinos in reactions where muons participate. If a detector, exposed to these neutrinos, only reports muons but no electrons, we know that muon neutrinos can only interact with muons.[10] This sounds simple, but in reality requires a lot of ingenuity to make it happen and a large new accelerator to boot. Let's see how it's done.

Leon Lederman, Jack Steinberger, and Melvin Schwartz[11] directed protons, accelerated in the AGS, onto a beryllium target, shown in the upper-right in Fig. 6.4. This produced copious numbers of pions, which decay within a few meters into muon neutrinos and muons.[12] A negative pion decays into a negative muon and its anti-neutrino,[13] labeled $\bar{\nu}_\mu$, whereas a positive pion decays into a positive muon and its neutrino. These processes are shown in respective encircled inserts on the top left and top right in the figure. Any not-decayed pions are readily absorbed in the upper layers of the shielding, but stopping the muons required several thousand tons of iron armor from a dismantled battleship. Only neutrinos make it through and continued towards

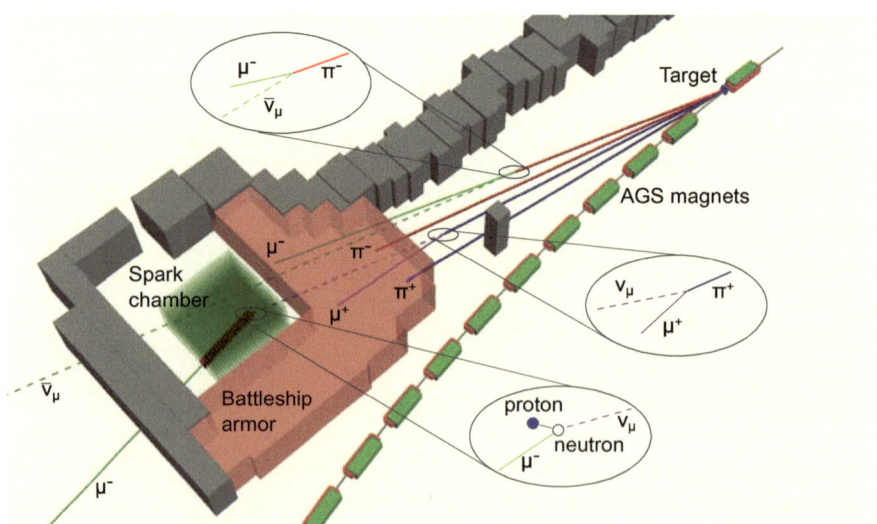

Fig. 6.4 Two-neutrino experiment

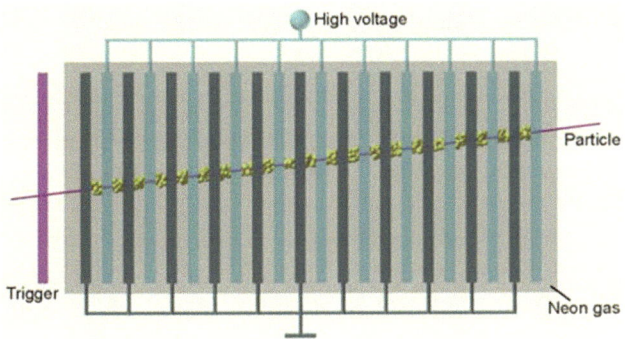

Fig. 6.5 Spark chamber

the detector, called *spark chamber*.[14] The anti-neutrinos do not interact in the detector, but a muon neutrino, when colliding with a neutron, produces a proton and a negative muon, as shown in the encircled insert on the lower right in Fig. 6.4. This negative muon was then identified in the spark chamber that is shown in Fig. 6.5. This detector was constructed as a sandwich of aluminum plates on high voltage, while the space between the plates is filled with neon gas. Any charged particle, passing through this sandwich, knocks out electrons from the neon gas. Just at this time the high voltage is turned on and a spark develops at the place where the free knocked-out electrons hang out, which leads to visible tracks on photographs. Of course tremendous ingenuity was needed to distinguish electron tracks from muon tracks. Moreover, the experimenters had to filter out unwanted events such as cosmic radiation producing events unrelated to the protons hitting the beryllium target.

The signature they looked for was a sequence of sparks between plates that looks like the track of a particle that starts in the middle of the detector and therefore had to be initiated by a neutrino. They had figured out how to differentiate tracks of electrons from those of muons and, incidentally, they only found muon tracks, which proved that muon neutrinos cannot produce electrons and therefore must be different from electron neutrinos. This is yet another piece of the puzzle on our way to complete the standard model: there are two types of neutrinos, one for electrons and one for muons.

Not enough with two types of neutrinos, the weak interaction had more surprises in store, this time caused by neutral kaons.

Kaons and CP Violation

In the summer of 1964 Val Fitch and James Cronin[15] decided to figure out some strange behavior of neutral K^0 mesons. They sometimes decay into two and sometimes into three pions.[16] The one that decayed into two pions did so after traveling a few cm and was denoted K-short. The one that decayed into three pions traveled a few meters before doing so and was denoted K-long. Fitch and Cronin therefore expected to only find K-long after several meters but, instead, found several hundred two-pion decays. How could that be?

It turned out that, despite being neutral, the K^0 and its anti-particle $\bar{K}^0$ are different particles. Moreover, owing to the quirky nature of quantum mechanics, the K-long and K-short are mixtures of both and can spontaneously appear as one or the other. These particles and their anti-particles can nilly-willy exchange their identities. That the mixture is slightly imbalanced was the real revolution; it favors matter over antimatter and that contradicted the deep-felt belief in the symmetry between matter and antimatter that physicists had at the time. This was almost as much a blow to their belief as Yang and Lee's "fall of parity" was. Cronin and Fitch's experiment gave the first indication as to why there is mostly matter in our universe, rather than antimatter, a fact that, for technical reasons, is called *CP violation*. The weak force appears to be rather liberal in messing with the symmetries of nature; both parity and CP are violated.[17]

But what about the strong force? Are there more surprises in store? More and more resonances and other new particles showed up in experiments. Hopes ran high to find some order in this "zoo of particles."

Eightfold Way

Roughly simultaneously with the proliferation of new particles and resonances,[18] Murray Gell-Mann and Yuval Ne'eman independently developed classification schemes[19] based on Group theory, a relatively abstract branch of mathematics that deals with transformations. As an example, consider a regular six-faced dice and assume that the number facing upwards describes the state of the dice. As transformations we consider a rotation of the dice by 90° around one of its three axes. Clearly, not all faces can be moved to face upwards by a single rotation; the one at the bottom requires two. In much the same spirit Gell-Mann and Ne'eman used a group called $SU(3)$ to describe the transitions from one particle to another. In order to visualize their idea, we display the particles on a diagram where the horizontal axis labels the particle's

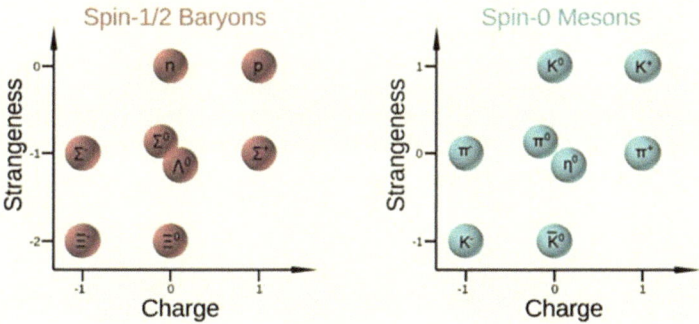

Fig. 6.6 Baryons with spin 1/2 (left) and mesons with spin zero (right)

charge and the vertical axis labels its strangeness. On this diagram, the spin-1/2 particles heavier than protons—the baryons—form the regular pattern, shown on the left-hand side in Fig. 6.6. The group theoretical considerations, for example, predict that there are two neutral particles with one negative unit of strangeness, the Λ^0 and the Σ^0. The analysis also predicted that there is no positively charged particle with two negative units of strangeness. Moreover, none was ever found in experiments.

Also the spin-0 particles that are lighter than protons—the mesons—among them the pions and the kaons follow similar group theoretical considerations. The right-hand side in Fig. 6.6 shows these particles depicted in the charge-strangeness diagram. All of them have properties that are consistent with experiments. At the time the model was constructed, the eta (η) meson was not yet known, but it was discovered at little later with the help of a pion beam extracted from the Bevatron. Note that the K^- as the anti-particle of the K^+ has opposite-sign charge and strangeness of the K^+ and likewise the K_0 and $\bar{K}^0$.

Even the properties of the heavier baryons with spin 3/2 fit into the group-theoretical framework shown in Fig. 6.7. The top row is populated by the four different charge states of the Delta-resonance and the two rows below with the three Σ^* and Ξ^*. Both are excited states with higher spin of the corresponding states shown on the left-hand side in Fig. 6.6. In particle-physics circles the story about the prediction of the last missing particle in the so-called decuplet is well-known: at a conference in the summer of 1962 Nicholas Samios reported evidence for the three Σ^* and the two Ξ^* in experiments with a negative kaon beam produced by an internal target at the AGS and directed into their bubble chamber. Upon hearing the news, both Gell-Mann and Ne'eman raised hands to comment. Gell-Mann, being better known, was given the word and he predicted the existence of a new particle and even its mass. At the same time

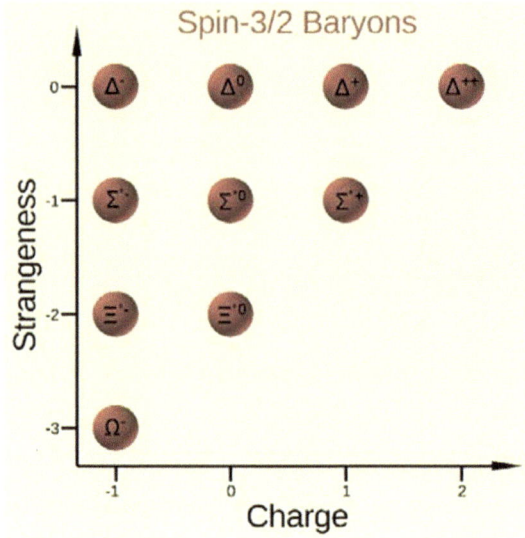

Fig. 6.7   Baryons with spin 3/2

he named it Omega-minus and gave it the symbol Ω^-. After the meeting, Gell-Mann and Samios met in a cafeteria and discussed how to find the new particle. On what must be one of the most famous napkins in the history of particle physics, Gell-Mann sketched the most favored reaction to detect the Omega-minus.

Omega-Minus

And Samios set out to find the Ω^-. He used a magnetic guidance system to isolate negative kaons, coming from an internal target in the AGS. He directed them to the recently commissioned large 80-inch hydrogen bubble chamber, where a large number of pictures was recorded.[20] Many showed interesting events, but the first Ω^- only showed up on picture 97025.

Figure 6.8 shows a cleaned-up rendition of the tracks in that event.[21] Once an interesting picture was identified, the curvature and thickness of all visible tracks were carefully measured and reconciled with known particle types. The analysis takes into account their masses, quantum numbers, as well as the conservation of energy and momentum of all participating particles. This includes the neutral particles that do not cause visible tracks. The "solution" to this puzzle from Fig. 6.8 is the following story. The negative kaon enters from the bottom before it hits a proton and immediately gives rise to an Ω^-, a K^0, and

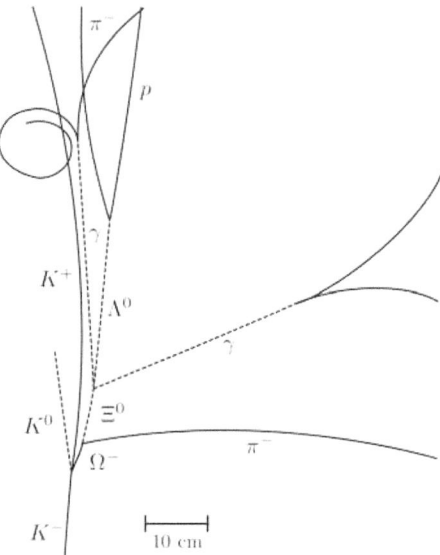

Fig. 6.8 The Omega minus (Public domain image from Wikimedia)

a K^+, which takes off to the top left. Note that the neutral K^0 does not leave a track and its existence is inferred in the process of solving the puzzle. After a short time, the Ω^- decays into π^- and Ξ^0. These in turn decay into the neutral (and therefore invisible) particles Λ^0 and π^0, which immediately decay into two photons, labeled γ in Fig. 6.8, each of which produces oppositely curling electron-positron pairs. The Λ^0 moves towards the top of the image and decays into proton and π^-, visible in the upper half of the plot. Within a few months, three similar images were identified, after carefully cross-checking all conservation laws, whose only feasible interpretation was that an Ω^- was involved.

While the predictions of the eightfold way were so spectacularly proven correct, it was still surprising that the rather abstract theory worked so well. Or is there some deeper order responsible for its success?

Quarks

So, how can we find a deeper order or maybe even more fundamental entities? Both Gell-Mann and George Zweig independently came up with a scheme to construct, for example, all baryons in Fig. 6.7, based on some underlying mathematical entities that Gell-Mann called *quarks*.[22] Both horizontally and vertically there are four baryons with three steps in between. Let's therefore

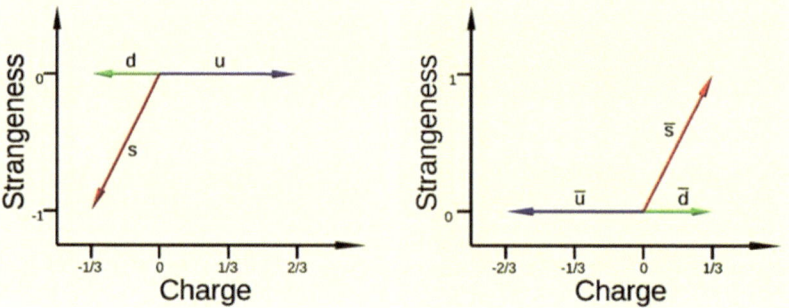

Fig. 6.9   Three quarks (left) and the corresponding antiquarks (right)

assume that we need three types of quarks. If we start at charge and strangeness zero, which is where the Δ^0 is located, reaching the Δ^- requires only one step to the left and no step downward. In order to get there with three copies of a quark, it better had charge $-1/3$ and strangeness 0. Since we go down in charge, let us therefore call this the *down* or *d* quark. We therefore hypothesize that the Δ^- is composed of three down quarks. Likewise, to take us from the origin of the diagram towards the right and upwards in charge, to the Δ^{++}, we need three copies of an *up*, or *u* quark with charge $+2/3$ and no strangeness. The Δ^{++} is thus composed of three *up* quarks. Finally to take us from the origin to the Ω^-, we need to go left by one unit and reduce the strangeness by three units. Three copies of the *strange*, or *s*-quark, will achieve just that. The Ω^- is thus composed of three *strange* quarks.

The left-hand diagram in Fig. 6.9 shows the solution of the puzzle that Gell-Mann and Zweig solved and that "explains" all spin-3/2 baryons in terms of quarks. The three quarks are illustrated as arrows in the same coordinate system already used in Fig. 6.7. Now we need to figure out how to assemble remaining baryons as combinations of these quarks. For example, each of the three quarks in the combination *dds* has charge $-1/3$, such that their combined charge adds up to -1 and the single strange quark makes the strangeness -1, which indicates that the Σ^{*-} contains the two *d* quarks and one *s* quark. Try out a few other combinations and determine the quark combinations that give the other members of the decuplet from Fig. 6.7.

Remarkably, it is also possible to find quark combinations that give us all the spin-1/2 baryons on the left-hand diagram in Fig. 6.6. For example, the neutron has the quark composition *udd*, the proton has the composition *uud*, and the Ξ^- has *ssd*.

But which ever way we try to assemble three quarks, the mesons from the right-hand side in Fig. 6.6 never come out right, because we have no quark

Fig. 6.10 Baryon and meson

to increase the strangeness. For that we need *antiquarks*. They are represented by arrows pointing in the opposite direction, as illustrated on the right-hand diagram in Fig. 6.9. Now we can construct all spin-0 mesons from the right-hand diagram in Fig. 6.6 as compositions of one quark and one antiquark. For example the K^+ is composed of one $\bar{s}$ antiquark to increase the strangeness to $+1$ and contributes $+1/3$ units of charge. The additional u quark contributes $+2/3$ units of charge. We thus obtain the K^+. Reversing the arrows, which is equivalent to replacing quarks by their antiquarks, result in the K^-, which has the composition $\bar{u}s$.

All of a sudden, the somewhat arbitrary distinction of baryons and mesons makes sense. The heavier baryons, shown on the left-hand side in Fig. 6.10, consist of three quarks and the usually lighter mesons, shown on the right-hand side, consist of a quark and an antiquark. Jointly, all particles consisting of quarks, both mesons and baryons, are collectively referred to as *hadrons*. Notable exceptions to hadrons are electrons, muons and neutrinos. They are referred to as *leptons*. The classification schemes of particles that were originally introduced somewhat ad-hoc, can now be consistently "explained" in terms of quarks and leptons.

Let us explore the relation between leptons and quarks some more. After all pions are hadrons, and we saw them in Fig. 6.4 decay into muons and neutrinos, which are both leptons. It turns out that the weak interaction is responsible.

Electroweak Unification

Already in the early 1930s, Enrico Fermi formulated a theory for the simplest radioactive process, that of a neutron decaying into a proton, an electron, and a neutrino (Chap. 3). Fermi's theory, however, had a problem. At the large energies the new accelerators were approaching, it would fail because of internal mathematical inconsistencies.

Theoreticians had previously experienced similar inconsistencies when creating a quantum theory for the interaction of charged particles with electromagnetic fields, called *quantum electrodynamics* (QED). But Richard Feynman,[23] Julian Schwinger, and Shin'ichiro Tomanaga[24] found a systematic way, called *renormalization*, to fix the inherent inconsistencies in QED. The crucial ingredient of their fix was that the theory could be formulated in terms of a symmetry group, called $U(1)$. Such symmetries are referred to as *gauge symmetries* with $U(1)$ being the corresponding *gauge group*. This group is a close relative to the $SU(3)$ group underlying the eightfold way to classify the mesons and baryons, but used in a different way. In QED, $U(1)$ describes an internal symmetry, in particular, it assumes that the wave function describing the state of a particle can be multiplied by a phase factor—an element of the group $U(1)$—without changing any experimental predictions derived from the theory.[25] This only works, if we simultaneously introduce some compensating quantities, which miraculously turn out to be the electro-magnetic fields that obey corresponding transformation laws. On the quantum level electro-magnetic fields appear as photons. They tell one charged particle that another one is nearby, but this only works if the particles are charged. Photons cannot "see" neutral particles. Photons thus turn out to be the carriers of the electro-magnetic force. Admittedly, deriving the whole theory from a symmetry requirement sounds a bit like magic, but it works! QED with the renormalization recipe predicts all experiments within the achievable measurement precision.

The success of introducing the phase factor in QED led Chen-Ning Yang—of parity-toppling fame—and Robert Mills to develop the mathematical framework to describe field theories based on other groups, especially $SU(N)$, which they published in 1954. Buried in their mathematical framework, the number of force carriers corresponding to the classical fields is $N^2 - 1$, such that for $N = 2$ this predicts three force carriers, and for $N = 3$ it predicts eight force carriers. In this framework all force carriers turned out to have one unit of spin. Particles with integral spin are collectively called *bosons*. They have the remarkable property that you can stuff many of them into the same spot; the many photons in a strong laser beam may serve as one illustration; the blue balls, all assembling on the ground floor on the right-hand side in Fig. 6.11 is another. In contrast, electrons and protons have half-integer spin and such particles are collectively referred to as *fermions*. They refuse to be in the same spot with other fermions of the same type because they obey Pauli's exclusion principle (Chap. 3). All particles that make up matter are fermions and cannot sit in the same spot, just as two travelers on a bus cannot occupy the same seat, and the red balls on the left-hand side in Fig. 6.11 must spread out over several floors.

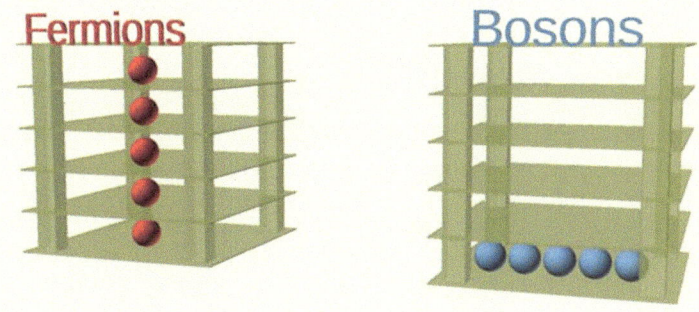

Fig. 6.11 Fermions and bosons

The distinction into fermions and bosons provides the key to the understanding of the fundamental forces. The force-carrying bosons communicating the presence of one fermion to another fermion, but only if the fermions carry the charge of the corresponding force. For the electromagnetic interaction the boson is the photon and the charge is the well-known electric charge. In Chap. 7 we will encounter gluons as the bosons of the strong force and "color" as the corresponding charge. The bosons responsible for the weak interaction are an integral part of a theory that emerged early in the 1960s. At the time Sheldon Glashow, Steven Weinberg, Abdus Salam,[26] and others applied the theory of Yang and Mills to the weak interaction. In order to account for electrons and neutrinos they suspected that the gauge group $SU(2)$ is part of the story. In order to also take QED into account they folded in the group $U(1)$, making their theory depend on both groups. It is commonly referred to as $U(1) \times SU(2)$. For the number of force carrying bosons, the theory thus predicted four, one from the $U(1)$ part—the photon—and three from the $SU(2)$ part, one neutral called Z^0, and two charged ones with opposite polarity, called W^+ and W^-.[27]

So, the hunt to find these new bosons of the weak interaction was on. One of the striking features in this theory is the neutral Z^0, which, using the jargon of the time, represents a *neutral current*. But to see it in action, it turned out that lots of neutrinos are needed.

Great Idea: Magnetic Horn

Soon after the CERN PS came into operation and the successful proof from the AGS that there are two types of neutrinos, the experimenters at CERN were eager to explore the reactions that the muon neutrinos could induce. But that required many more muon neutrinos than the PS could produce all by

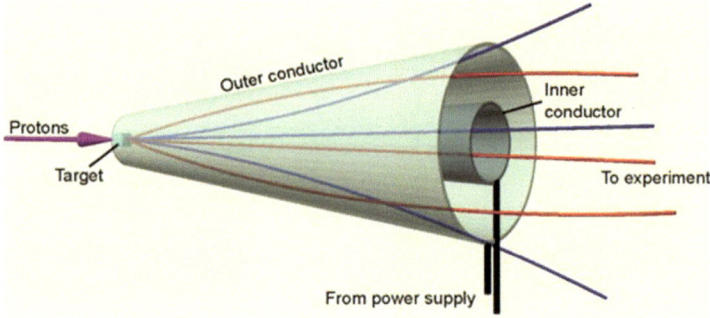

Fig. 6.12   Magnetic horn

itself. Simon van der Meer[28] and colleagues therefore analyzed the limitations
of the process in which the muon neutrinos are created in the decay of pions
into muons and neutrinos. They noticed that many of the pions, which are
produced by smashing protons into a target, escape at large angles, rather than
nicely moving in the forward direction towards the experiments. In order to
bend these escaping pions back in the forward direction, they invented the
magnetic horn made of the two concentric hollow pipes shown in Fig. 6.12.
Huge electric currents, provided by a discharging capacitor bank, rush in one
direction through the inner pipe and in the opposite direction through the
outer pipe. In this way a strong magnetic field develops in the space between
the pipes that bends the pions of one polarity (red) back to the forward direction
giving their decay products, muons and muon neutrinos, a better chance to
reach the experiments. On the other hand, pions of the opposite polarity (blue)
are defocused and spread out such that only very few end up moving towards
the experiment. Depending on the polarity of the current, the horn either
focuses positive or negative pions. The opposite polarity defocuses and takes
the opposite-polarity pions to never-never land. Since positive pions decay into
positive muons and muon neutrinos, and negative pions into negative muons
and muon antineutrinos, the direction of the current selects which type of
neutrino reaches the experiments.

This device proved decisive to significantly increase the flux of neutrinos
and made it possible to figure out whether neutral currents as a manifestation
of the Z^0 really exist.

Weak Neutral Currents

By 1970 CERN was the home of the huge bubble chamber *Gargamelle*. With a length of almost 5 m it held 12 cubic meters of liquid to observe neutrino-initiated reactions. Two types of these reactions were identified as the "smoking gun" signatures of neutral currents.

Without neutral currents, muon neutrinos need a muon to interact with electrons. Conversely, neutral currents, the Z^0s, are needed for a muon neutrino to kick an electron without a muon participating in the interaction. The experimenters therefore tried to find reactions in their bubble chamber where an electron spontaneously starts moving with high speed at a small angle without any muons nearby. Only neutral currents, or Z^0s, can explain such a reaction. In the second type of reaction, the muon neutrino hits a nucleus and creates lots of escaping hadrons. Without a simultaneously appearing muon such an event is only possible if neutral currents are involved. The experimental signature is thus a shower of hadrons that spontaneously pops out of nowhere.

Both types of reactions were observed in Gargamelle. A particularly tricky part of the analysis was to rule out causes, other than muon neutrinos, that could trigger the observed signature. Diligent detective work made sure that the observed events were indeed caused by neutral currents,[29] or in other words, by a Z^0, though the particle itself was not directly observed. Gargamelle only provided indirect evidence of its existence. The direct observation of Z^0s came later.

Gargamelle was the last of the great bubble chamber. Extracting information from pictures was just too slow; a new computer-readable detection device was needed.

Great Idea: Multi-wire Proportional Chamber

Already in the late 1960s, Georges Charpak[30] developed new types of detectors that could be automatically read out from a computer, the *multiwire proportional chamber* shown in Fig. 6.13 and the *drift chamber*. Both are based on placing a large number of wires at positive high voltage and immersed in a gas between plates at negative voltage. A passing particle knocks out electrons from the gas such that they and the remaining ions move towards the electrodes and create an electric pulse. The signal on the wire that is closer to the high-energy particle carries a larger signal and the signal comes a little earlier. By carefully analyzing the motion of both electrons and ions Charpak was able to dramatically improve the resolution to determine the trajectory of the

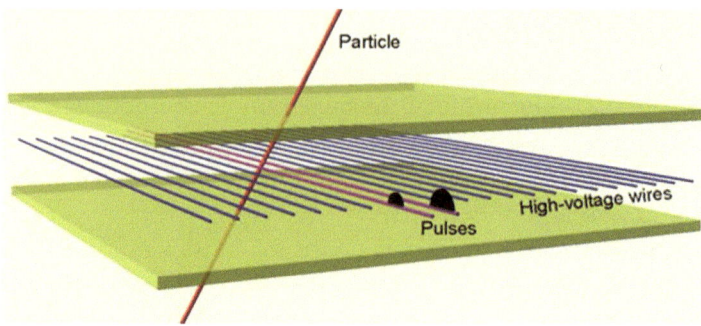

Fig. 6.13   Multi-wire proportional chamber

originating particle. Practically all detectors of charged particles used ever since are based on Charpak's discoveries.

Particles and Forces

Slowly a picture of the subatomic world, shown on the inside of the front cover, emerged.[31] All matter consists of either quarks or of leptons—electrons, muons, or neutrinos. All of them are fermions. As a consequence, they obey Pauli's exclusion principle and refuse to share the same spot, actually the quantum-mechanical state, with another fermion of the same type.

And these matter particles interact with each other by sending force-carrying bosons, such as photons, Z^0, and maybe pions back and forth. Bosons do not obey the exclusion principle and can pile up. This is how forces become more and more intense.

But at the time, during the mid-1960s, several mysteries remained unsolved. It was unclear whether pions are really the force carriers of the strong nuclear force. That nobody had any experimental evidence for the existence of quarks was another headache. It was time to lift the veil over these mysteries. And that required a new type of accelerator.

Notes

1. The early years are described in: John Kriege, *I. I. Rabi and the Birth of CERN,* Physics Today, September 2004, page 44. For a more detailed account of this era, see: Robert Jungk, *The Big Machine,* Charles Scribner's Sons, New York, 1968.
2. The travel report of the CERN delegation is available from https://cds.cern.ch/record/24911.

3. This is explained by one of the inventors in: Ernest Courant, *A 100-Billion-Volt Accelerator,* Scientific American, May 1953, page 40.

4. The idea of alternating-gradient focusing was discovered a few years earlier by the Greek engineer Nicholas Christofilos. Rather than publishing his idea in a scientific journal, he filed for a patent that went unnoticed. After this was later realized, Livingston and colleagues acknowledged his priority and offered him a job to work in Brookhaven. He accepted and worked there for a number of years before moving on to work on nuclear fusion.

5. After working on the first electron ring where synchrotron light was detected (Chaps. 8 and 13), John Blewett (1910–2000) moved to Brookhaven, where he played a leading role in the construction of the Cosmotron (Chap. 5), the AGS, and the new CERN PS.

6. For an overview see: Robert Wilson, *Particle Accelerators,* Scientific American, March 1958, page 64.

7. A vivid account of this night is given by Hildred Blewett (1911–2004) in https://cerncourier.com/a/a-night-to-remember/.

8. The early history of BNL is briefly told in: Ernest Courant: *Early History of the Cosmotron and AGS at Brookhaven,* BNL Report 36659, 1985. Available online from https://inis.iaea.org/search/search.aspx?orig_q=reportnumber:BNL--36659.

9. Frederik Reines (1918–1998) received the Nobel Prize in Physics in 1995 "for the detection of the neutrino and for pioneering experimental contributions to lepton physics" in both his and Cowen's name. Clyde Cowan (1919–1974) had sadly passed away much earlier.

10. If both muons and electrons came out this would indicate that the muon neutrino behaves just like the electron neutrino and interacts with both muons and electrons.

11. Leon Lederman, Jack Steinberger (1921–2020), and Melvin Schwartz (1932–2006) received the Nobel Prize in Physics in 1988 "for the neutrino beam method and the demonstration of the doublet structure of the leptons through the discovery of the muon neutrino."

12. For a first-hand description of the experiment see: Leon Lederman, *The Two-Neutrino Experiment,* Scientific American, March 1963, page 60.

13. We will denote anti-particles by a bar over the symbol.

14. Gerard O'Neill, *The Spark Chamber,* Scientific American, August 1962, page 36.

15. Val Fitch (1923–2015) and James Cronin (1931–2016) received the Nobel Prize in Physics in 1980 "for the discovery of violations of fundamental symmetry principles in the decay of neutral K-mesons."

16. This process involving neutral K^0 mesons is different from the one involving charged K^+ mesons in the tau-theta puzzle from Chap. 5.

17. For a more detailed account see: Eugene Wigner, *Violations of Symmetry in Physics,* Scientific American, December 1965, page 28. Eugene Wigner (1902–1995) received the Nobel Prize for Physics in 1963 "for his contributions to the theory of the atomic nucleus and the elementary particles, particularly through the discovery and application of fundamental symmetry principle."

18. An account of the ideas to restore order in the subatomic world see: Geoffrey Chew, Murray Gell-Mann, and Arthur Rosenfeld, *Strongly Interacting Particles,* Scientific American, February 1964, page 74.
19. Gell-Mann gave this classification scheme the whimsical name *eightfold way,* probably because the number 'eight' plays a role in the mathematical framework based on the group $SU(3)$ and is an allusion to the eightfold path in Buddhism.
20. A first-hand account of the experiment is given in: William Fowler and Nicholas Samios: *The Omega-minus Experiment,* Scientific American, October 1964, page 36.
21. Considering that someone looked at all 97024 pictures before gives us an idea of the enormous task that scanning bubble-chamber pictures was.
22. Zweig had called them *aces,* but Gell-Mann's chosen name *quarks* prevailed. He tells the origin of his choice in: Murray Gell-Mann *The Quark and the Jaguar,* Abacus, London, 1995, page 180. Apparently he wanted a new word for a new concept and then hit upon the line "Three quarks for muster mark" while perusing James Joyce's novel *Finnegan's Wake.* See also: https://www.sciencefriday.com/articles/the-origin-of-the-word-quark/.
23. Feynman's many contributions to physics and other fields are described in: James Gleick, *Genius,* Vintage books, New York, 1993. A very entertaining sort-of autobiography is: Richard Feynman, *Surely you're joking Mr. Feynman,* Bantam Books, New York, 1986.
24. Richard Feynman (1918–1988), Julian Schwinger (1918–1994), and Shin'ichiro Tomanaga (1906–1979) received the Nobel Prize in Physics in 1965 "for their fundamental work in quantum electrodynamics (QED), with deep-ploughing consequences for the physics of elementary particles."
25. The state of a particle is described by its wavefunction and the phase factor describes a rotation in an abstract mathematical space. Moreover, making the phase factor depend on space and time requires additional mathematical functions in order to make the theory mathematically consistent. Remarkably, these additional functions are related to the electric and magnetic fields. Since all experimentally observable quantities depend on the magnitude of the wavefunction, they are oblivious to the phase factor, but the electro-magnetic fields appear as a byproduct. Theories that are constructed by introducing generalized rotations that leaves the observable quantities unaffected, but introduce additional fields as byproducts, are referred to as *gauge theories.*
26. Sheldon Glashow (b. 1932), Steven Weinberg (1933–2021), and Abdus Salam (1926–1996) received the Nobel Prize in Physics in 1979 "for their contributions to the theory of the unified weak and electromagnetic interaction between elementary particles, including, inter alia, the prediction of the weak neutral current."
27. The description is somewhat simplified. Actually, the photon and the Z^0 are mixtures of the mathematical fields of the theory.
28. Apart from the magnetic horn, Simon van der Meer (1925–2011) also invented stochastic cooling (Chap. 9) and a new method to ensure head-on collisions of micron-sized beams. Jointly with Carlo Rubbia he recieved the Nobel Prize in

Physics in 1984 "for their decisive contributions to the large project, which led to the discovery of the field particles W and Z, communicators of weak interaction."

29. It was not easy to convince the physics community, because a group at Fermilab was hunting neutral currents as well. They used the 300 GeV proton beam from the main ring (Chap. 9) and, after a preliminary analysis, announced their discovery. After a more thorough analysis, however, the neutral currents disappeared from their data. This, in turn, made the CERN physicists worry about the validity of their result, but after carefully reanalyzing their data, they stood their ground and published their result. After the Fermilab group rebuilt their detector, the neutral currents even reappeared in their data, which caused the community to joke about *alternating neutral currents*. An account of what happened at the time is given by David Perkins in https://cerncourier.com/a/neutral-currents.

30. Georges Charpak (1924–2010) received the Nobel Prize in Physics in 1992 "for his invention and development of particle detectors, in particular the multiwire proportional chamber." He explains the physics in: Georges Charpak: *Multiwire and Drift Proportional Chamber,* Physics Today, October 1978, page 23.

31. First-hand accounts by two of the main actors shaping this view can be found in: Steven Weinberg, *Unified Theories of Elementary-Particle Interactions,* Scientific American, July 1977, page 50, and *The Forces of Nature,* Bulletin of the American Academy of Arts and Sciences January 1976, page 13, online available from https://www.jstor.org/stable/3823787. Sheldon Glashow with Ben Bova, *Interactions, a journey through the mind of a particle physicist and the matter of this world,* Warner Books, New York, 1988.

7

A Monster Encounters Quarks

The great advantage of the proton synchrotrons are the high energies of the accelerated particles, which were instrumental in populating the zoo of particles. And quarks seemed to play an essential role to establish order in that zoo. But nobody had seen isolated quarks, despite many ingenious experiments to find them. They seemed to forever hide inside protons and neutrons. To look there, an new type of accelerator was needed. It should provide a beam of, preferably point-like, particles. Electrons, accelerated to high energies, actually fit the bill. They have short enough quantum-mechanical wavelengths to actually "see" what goes on inside protons. But to accelerate electrons to very high energies requires a lot of power.

In an early development, already in 1934, William Hansen[1] at Stanford University thought about improving X-ray sources and he needed high-power radio-frequency waves with short wavelengths, so-called microwaves, to accelerate electrons before impinging them on a target to make X-rays.[2] At about the same time Sigurd and Russel Varian[3] were interested in improving aircraft guidance systems, both to detect aircraft from the ground and to detect the ground from inside the aircraft which would allow aircraft to land in bad weather. Also they came to the conclusion that high-power microwaves are needed for the job. Unaware of classified research towards *radar* in the US and in Britain, they decided to develop their own source of microwaves, the *klystron*.

© The Author(s), under exclusive license to Springer Nature Switzerland AG 2024
V. Ziemann, *Beams*, Copernicus Books,
https://doi.org/10.1007/978-3-031-51852-2_7

Great Idea: Klystrons

After several unsuccessful attempts Russel Varian asked his old college buddy Hansen, who had become a professor by the time, for help. Together they talked the head of the Stanford physics department into supporting the Varian brothers with 100 $ and access to the workshops, in exchange for a 50 % share of all future profits coming from their invention. Considering that *Varian Associates* over time became a billion-dollar enterprise, this turned out to be a very profitable investment.

It did not take too long until Russel realized that using a small cavity excited by microwaves would modulate the velocity of an electron beam.[4] Figure 7.1 illustrates this idea. A high-intensity electron beam, accelerated towards the ring-shaped positive electrode, has its velocity modulated by an electric field in the input cavity. Once the fast electrons catch up with the slow ones, the electrons are bunched in small packages in a process called *velocity bunching*. The bunches then induce a much higher field in the output cavity, from where radio-frequency power is extracted. A small input signal is thus amplified to a large output signal.[5] Klystrons for moderate power levels fit on aircraft and played a decisive role in radar applications during the Second World War, which Hansen helped to develop.

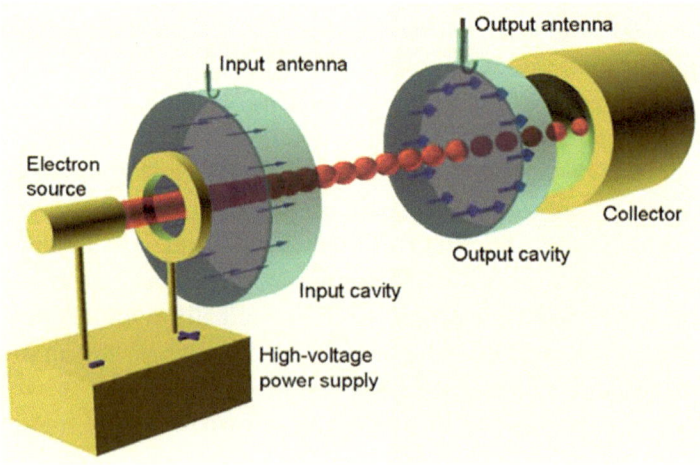

Fig. 7.1 Klystron

Only after the war could Hansen devote his attention to using klystrons for high-energy electron accelerators. The first challenge was to transfer the power from the microwaves to the electrons. He came up with an ingenious idea.

Great Idea: Disk-Loaded Waveguides

In order to transfer the power from the microwaves to the electrons he had to confine both in a narrow metallic pipe, a so-called waveguide.[6] The problem now is that the electrons always move slower than the microwaves which would pass over the electrons. The electrons would sometimes see accelerating and sometimes decelerating fields which, over time, would average to zero. To prevent this from happening Hansen slowed down the microwaves by inserting washer disks in the waveguide, as shown in Fig. 7.2. The dimensions of the disks are chosen such that the electric field of the microwave, shown by blue arrows, points in opposite direction between adjacent disks. Electrons, shown as the red ellipsoids, that arrive at the time when the field in the first "cell" points in the right direction, is accelerated. A little time later, it arrives in the next cell. But by that time the field has reversed polarity and the electron is accelerated once more. Unlucky electrons that arrive at the "wrong" time are decelerated. As a consequence such *disk-loaded waveguides* can only accelerate electrons in small packages, a few millimeters long and called *bunches*, that arrive at just the right time.

With the power transfer from microwave to electrons under control, Hansen was ready to validate his ideas by building the first electron linear accelerator, or *linac* for short.

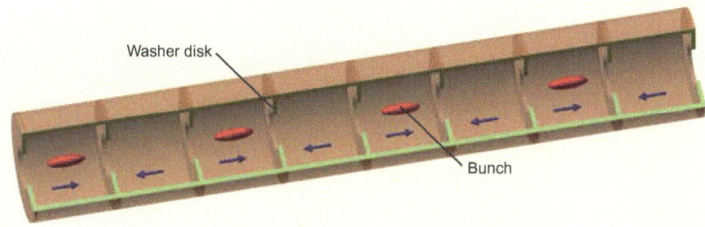

Fig. 7.2   Disk-loaded waveguide

Stanford Mark I to III Linear Accelerators

His first linac, named Mark I, was powered by a 900 kW magnetron, a predecessor of the klystron. It accelerated electrons in disk-loaded waveguides, also
called *accelerating structures*, with a diameter of 9 cm and a length of 90 cm to
an energy of 1.5 MeV. After installing additional waveguides he reached 6 MeV
in the now four-meter-long linac. The electrons reached the target in bursts
of closely spaced bunches, each burst having a duration of a little less than a
millionth of a second and did so sixty times per second. The achievable energy
was mostly limited by the available microwave power, but by the end of the
1940s much more powerful klystrons became available from the Sigurd and
Russel Varian's company *Varian Associates*.

And these new klystrons, reaching huge power levels of up to 20 MW in
$2\,\mu$s-long pulses, powered the Mark II linac, which accelerated electrons to
energies of 40 MeV. The Mark II was only a stepping stone along the way to
the self-declared goal to build a 1 GeV linac, the Mark III.

Like most of the later linacs the Mark III is constructed of 3 m-long accelerating structures with the same 9 cm diameter as its predecessors. Each structure
is powered by a single 20 MW klystron producing microwaves with a frequency
of 2.8 GHz.[7] In a first stage three structures brought the electron energy up
to 75 MeV. By 1954, installing additional accelerating structures brought the
energy up to 600 MeV. At this point the linac was almost 70 m long and delivered 10^{11} electrons sixty times per second. Later it was upgraded further and
could deliver electrons with 1.2 GeV, surpassing the original mission goal.

But already the 600 MeV electrons could probe the interior of protons.

The Size of Nuclei

Inspired by Rutherford's experiments with alpha particles and gold foils some
40 years earlier, Robert Hofstadter[8] directed high-energy electrons from the
Mark III linac at various targets where he hoped the electrons would resolve
details of the target nuclei. Like Rutherford, he detected the elastically scattered
particles, those that did not lose energy in the target, and counted how many
were deflected by a specific angle. The left-hand image in Fig. 7.3 illustrates the
process in which the electron interacts with a proton and softly bounces off,
leaving the proton unharmed. He therefore used the large spectrometer magnet,
shown in Fig. 7.4, to select the energy of the scattered electrons. Different
scattering angles are probed by rotating the magnet, which is mounted on a
salvaged gun-mount, around the target. At the top of the magnet the scattered

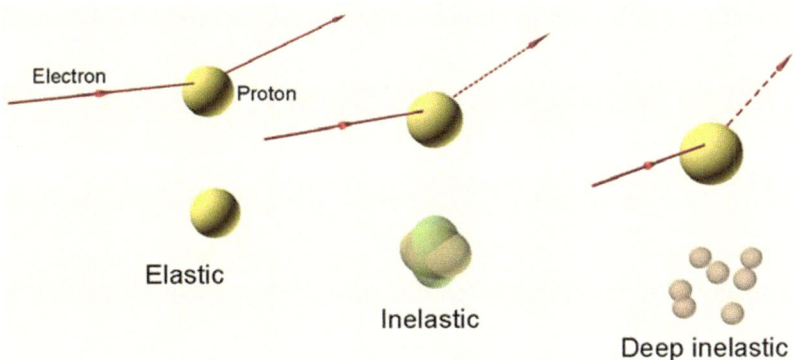

Fig. 7.3 Electron scattering from a proton: elastically (left); exciting a resonance (middle); deeply inelastic (right)

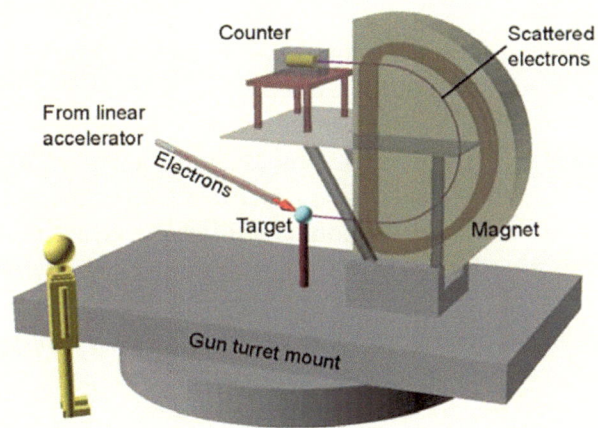

Fig. 7.4 Hofstadter's spectrometer

electrons with a specific energy intercept a scintillator and produce a flash of light that is counted with the help of some electronic circuitry.

Looking at the nucleus with much better resolution than Rutherford, Hofstadter found that the nucleus actually has a finite size. His scattering data agreed remarkably well with theoretical predictions. Even better, he could determine the shape of the nucleus, or more accurately, the charge distribution; as expected he found the maximum charge in the center, but its charge density decreased over a distance of about 10^{-15} m to zero outside. This was the first experiment that could look into the nucleus and measure some of its properties quantitatively and Hofstadter was honored with a Nobel prize in 1961.

Measuring the charge distribution of nuclei, even that of hydrogen—protons—is really cool, but what can we do to discern even smaller structures inside. The answer is, of course, to build a bigger accelerator.

Project M—The SLAC Linac

The success of the Mark III linac led a group of Stanford physicists, headed by Wolfgang "Pief" Panofsky,[9] to discuss their dream machine, the longest linac that would fit on university-owned land. They came up with a 3 km-long behemoth, called project M—for monster—that would bring electrons up to energies of 20 GeV. After negotiations with several national agencies, funding for the new national lab, the *Stanford Linear Accelerator Center*, or SLAC for short, was approved in 1961 and construction of the linac started soon afterwards.[10]

The SLAC linac is basically a longer version of the Mark III, though with improved hardware. All lessons learned from operating the Mark III and doing experiments were absorbed into its design. The linac is powered by 240 klystrons, housed in an above-ground building, the klystron gallery. The power is then transported to the accelerating structures, which are placed in a dug-out tunnel about 8 m below the klystrons.[11] Diagnostic elements along the linac determine the position and the shape of the beam. Furthermore, magnetic lenses, quadrupoles (Chap. 6), interspersed with the accelerating structures, ensure that the beam does not stray from its path and is well-focused when it is delivered to the experimental areas at the end of the linac. By the summer of love, in 1967, the experimenters at endstation A around Richard Taylor, Henry Kendall, and Jerry Friedmann[12] were eagerly expecting the first high-energy electrons.

In endstation A three large spectrometers, conceptually similar to the one shown in Fig. 7.4 and optimized for different energy ranges, determine the energies of electrons scattered from a target[13] at certain angles. These spectrometers could, however, at the same time, determine the elastically scattered electrons, shown on the left-hand side in Fig. 7.3, as well as those that lost energy in the target. With hydrogen—proton—targets, for example, this made it possible to detect excited states, or resonances, of the proton, which is illustrated in the center of Fig. 7.3. But something unexpected happened in reactions where the electron receives such a large recoil that the proton is blown into smithereens. This process is commonly referred to as *deep inelastic scattering* and is illustrated on the right-hand side in Fig. 7.3.

Deep Inelastic Scattering

While Hofstadter determined the size of nuclei, experiments done at proton accelerators showed that most of the protons directed at nuclei almost went straight through, suggesting that their interior is somewhat diffuse. So, everyone had expected that also electrons went straight through and that only very few of them would be scattered towards large angles. Contrary to these expectations, a substantial number of electrons were kicked to large angles and, at the same time, blew the proton into pieces. The large angles give an impression of *deja vu* from Rutherford, who found that large deflection angles of his alpha particles required a teeny-tiny nucleus inside atoms to explain the large angles. Now there seem to be even smaller, maybe even point-like, scattering centers deeply hidden inside the protons. It's kind of funny, using low-energy alphas and the nucleus looks point-like; using medium-energy electrons from the Mark III and the nucleus looks like a fuzzy ball of charge. Now, looking with very-high-energy electrons, all of a sudden, a substructure emerges from inside the nucleus.[14]

This was not unexpected to everyone. SLAC theoretician James Bjorken had worked out how high-energy electrons would interact with the protons in the target. He had used a theoretical framework called *current algebra* to determine, in particular, the fate of the scattered electrons; at what angles and energies they would show up after their violent encounter with the target, which is just what his colleagues at endstation A were measuring. Deeply buried inside the heavy mathematics the assumption was that scattering happens between point-like electrons and point-like "things" inside the proton. Here point-like has an important implication, because a point looks the same no matter from what distance we look at it. In other words, it has no intrinsic size that may serve as a yardstick. This absence of a yardstick implies that probing the interaction on different length scales, or equivalently different energies, yields essentially the same result. Bjorken called this behavior *scaling*. The problem was, that the experimenters did not understand what he was talking about. The math was just too far out. Yet, their measurements matched Bjorken's calculations.

In a serendipitous coincidence, Richard Feynman, visiting his sister who lived nearby, dropped in on colleagues at SLAC to talk shop. One of them showed him the scattering data superimposed on Bjorken's calculations and confided that they are puzzled by the agreement; maybe Feynman could help out with some intuition of what goes on inside protons.

Partons

At first Feynman could not quite see it either, but after a while the smoke screen lifted once he visualized himself riding on the electron and seeing the proton rushing towards him. Since the relative speed of proton and electron is close to the speed of light, Einstein's special relativity dictates that the proton is longitudinally compressed and appears as a pancake coming at Feynman. If the proton has a substructure, it would appear like nuts in the pancake. Feynman called these nuts *partons* and worked out how the electrons deflect from them. He found almost the same results Bjorken did, but in his framework Bjorken's plots obtained a tangible interpretation. They revealed the momentum distribution of the partons.

The strong visual image of partons whizzing inside protons and the consistency of experiment and theory convinced many physicists that the partons are real and not just mathematical artefacts. But many questions remained open. Determining the spin of the partons was high on the agenda. It could soon be extracted from the angular dependence of the scattered electrons. The data clearly favored that partons are spin-1/2 particles. At this point many physicists started wondering whether partons are the same as quarks, in which case Taylor, Kendall, and Friedmann had found quarks inside protons.

Additional evidence could come from scattering experiments with neutrons instead of protons, because they contain different mixtures of quarks. Gell-Mann and Zweig's quark model predicted quarks to have fractional electric charges: 2/3 for the *u* quarks and −1/3 for the *d* quark, which would imply that scattering electrons from protons with quark composition *uud* and from neutrons with composition *udd* would show a difference. The problem is that there is no target of isolated neutrons, because they radioactively decay after a short while. The next best thing is to use a target made of deuterium, having one proton and one neutron inside its nucleus, and subtract the results from that of protons alone. In this way only those events where the electron scatters from the neutron are counted. It turned out that the measured ratio of electrons scattered off protons and off neutrons was different, and the difference agreed with predictions from the quark model. The deep-inelastic scattering experiments thus amassed crucial evidence for the existence of quarks.[15]

Despite the success of the quark model, it was still disturbing that no free quark, or any other particles with fractional charge, had been detected outside protons or neutrons. If they are really partons and move freely inside protons, what keeps them from coming out? Moreover, adding up the momenta of the spin-1/2 partons only gave half the total momentum of the proton. There must be something else inside the proton.

Quantum Chromodynamics

Drawing on parallels to the successful description of the electroweak theory in terms of the $U(1) \times SU(2)$ gauge group discussed in Chap. 6, the gauge group $SU(3)$ emerged as a candidate to describe the strong interaction that governs the inside of nuclear particles. As a consequence, quarks would assume a new property (or quantum number), that was called *color charge*, or *color* for short. Instead of a single *u*-quark, now there is a red *u*, a green *u*, and a blue *u* quark. Note that red, green, and blue are merely labels to describe the different types of color charge; they have nothing to do with "color" in the everyday sense. Since the "3" in $SU(3)$ implies that there are three color charges, the analogy to the three base colors was just a convenient mnemonic. So, how do the quarks tell each other about their color? They send *gluons* around, in much the same way that electrically charged particles talk to each other by sending photons around. Moreover, rather than a single gluon, the math of $SU(3)$ dictates that there are eight different gluons needed to account for the chatter among quarks. And these gluons turned out to be responsible for the missing momentum inside protons. This emerging theory of the strong interaction was soon dubbed *quantum chromodynamics*, or QCD for short. There is, however, a crucial difference between photons and gluons. Photons do not carry electric charge whereas gluons carry color-charge. They can therefore interact among themselves. Even hypothetical particles, consisting of only gluons and called *glueballs*, are conceivable, though not yet convincingly found in experiments.

The scheme with the three colors gives an intuitive picture as to why baryons and mesons do not show color to the outside observer. Baryons are compounds of three quarks: one red, one green, and one blue add up white, just like the fundamental colors jointly produce white light. Likewise, a quark with some color and an antiquark with the corresponding anticolor give rise to mesons, which do not exhibit color to the outside either. Thus all particles found in experiments are colorless.

Color charge also solved an annoying problem with the Δ^{++} particle, which is composed of three *d*-quarks, all having spin 1/2, and are therefore fermions. But Pauli's exclusion principle forbids multiple fermions to occupy the same state. Giving the three quarks different colors makes them different and so they are actually allowed to occupy the same state. Additionally, without colored quarks, the life time of the neutral pion turned out to be much too long in calculations. Only after taking color into account did the calculated life time agree with the measured one. Add another check mark for QCD!

But there's more. A real jackpot.

Asymptotic Freedom

Around 1973, David Gross, Frank Wilczek, and David Politzer,[16] applied a mathematically ambitious technique, called *renormalization group analysis*, to the forces inside the proton and how they change with distance.[17]

In order to illustrate the idea, let us consider quantum electrodynamics first and imagine a single electron, which is, however, not alone. It is surrounded by a cloud of spontaneously appearing and vanishing photons and electron-positron pairs. As long as these pairs recombine quickly enough, this is actually permitted by the laws of quantum mechanics, and especially Heisenberg's uncertainty relation.[18] Particles that take out a short-term loan of energy and repay it quickly enough are called "virtual." If we look at an electron from far away this cloud of virtual electron-positron pairs screens the electron in the center. If we then move closer to the central electron, this screening is more and more reduced. In other words, the charge of the central electron increases as we approach it. Note that the virtual photons do not participate in the screening, because they carry no electric charge.

Now consider QCD, which differs from QED in a crucial point; the gluons do carry color charge. This implies that they contribute to the screening of a central quark. Gross, Wilczek, and Politzer found that the contribution of gluons overcompensates that of the virtual quark-antiquark pairs and that the color force of the central quark goes to zero as we approach it, a behavior that was dubbed *asymptotic freedom*. Figure 7.5 illustrates this by the weak springs linking the quarks on the left-hand side. The springs get thicker and stiffer as

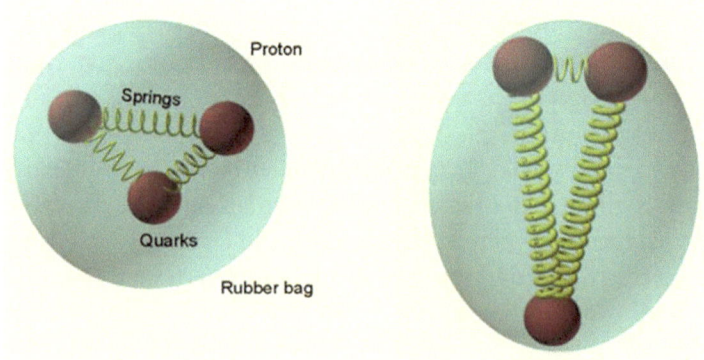

Fig. 7.5 Asymptotic freedom

the quarks are pulled apart, which is the situation shown on the right-hand side. It almost appears as if the quarks are captured in a rubber bag that stretches more and more as the quarks try to flee each other. But on the inside they behave like free particles, just as Feynman had assumed when he came up with the parton model.

Now we have an idea why quarks move freely inside protons and why they never come out. On the other hand, we do not know, whether there are other quarks besides u, d, and s. Great progress towards answering this question came from a new class of accelerators: electron rings.

Notes

1. William Hansen (1909–1949) was the driving force behind the early developments of linear accelerators at Stanford University.
2. The triode vacuum tubes available at the time could neither work at so high frequencies nor could they provide the required high power levels.
3. Sigurd (1901–1961) and Russel Varian (1898–1961) were founders of Varian Associates. Their company, being one of the early high-tech companies in Silicon Valley, grew from the initial 100 $ investment to a billion-dollar enterprise.
4. Edward Ginzton (1915–1998), a former student of Hansen, led the linac developments at Stanford University after Hansen's death. He gives a first-hand description of the developments in: Edward Ginzton, *The Klystron,* Scientific American, March 1954, page 84.
5. Taking a small part of the output and feeding it back to the input causes a continuous pile-up of microwave power to very high levels.
6. For a first-hand account see: Wolfgang Panofsky, *The Linear Accelerator,* Scientific American, October 1954, page 40.
7. This is close to the frequency used in modern microwave ovens, which, however, operate at 1000 times lower power levels—kW instead of MW.
8. Robert Hofstadter (1915–1990) received the Nobel prize in Physics in 1961 "for his pioneering studies of electron scattering in atomic nuclei and for his thereby achieved discoveries concerning the structure of the nucleons." A first-hand account can be found in: Robert Hofstadter, *The Atomic Nucleus,* Scientific American, July 1956, page 55.
9. After working with Lawrence in Berkeley, Wolfgang Panofsky (1919–2007) moved to Stanford and worked on the development of linear accelerators and storage rings (Chap. 8). He later became the founding director of SLAC.
10. For a first-hand account see: Edward Ginzton and William Kirk, *The Two-Mile Electron Accelerator,* Scientific American, November 1961, page 49.

11. The SLAC linac was actually in the way of the planned Highway 280 from San Francisco to San Jose. After some negotiations, the California Department of Highways agreed to construct a bridge passing over the klystron gallery ahead of schedule in order to prevent disturbances of running the delicate linac later. For several years a "bridge to nowhere" graced the landscape. By the early 1990s, the highway had materialized and traffic had increased significantly. Therefore its influence on the, by that time much more delicate, accelerator was a worry. A colleague and I, being post doctoral researchers at SLAC at that time, were tasked to measure the influence of traffic over the bridge on the accelerator. It turned out, however, that the noise induced by the highway, was low enough not to adversely affect operating the SLC (Chap. 10).

12. Richard Taylor (1929–2018), Henry Kendall (1926–1999), and Jerome Friedmann (b. 1930) received the Nobel Prize in Physics in 1990 "for their pioneering investigations concerning deep inelastic scattering of electrons on protons and bound neutrons, which have been of essential importance for the development of the quark model in particle physics."

13. Usually hydrogen, cooled to temperatures where it becomes liquid and has a higher density than the gas, was used.

14. A first-hand description is given in: Henry Kendall and Wolfgang Panofsky, *The Structure of the Proton and the Neutron,* Scientific American, June 1971, page 60.

15. The SLAC linac only bombarded targets 120 times per second and did so until the linac was reconfigured as a linear collider (Chap. 10). At this point other accelerators took over the business of deep-inelastic scattering. In order to explore very rare processes, a continuous stream of high-energy electrons is desirable and that is what the *Continuous Electron Beam Facility* (CEBAF) in Newport News provides since 1994. Two 240 m long superconducting linacs are connected by recirculating magnetic arcs, such that they can repeatedly accelerate a continuous beam of electrons to 6 GeV, and later 12 GeV. Several experiments in a number of endstations explore the constituents of nuclei with high precision. The *Hadron Electron Ring Accelerator* (HERA) at DESY extended the energy reach of the collisions with a 27.5 GeV electron beam and a 920 GeV proton beam. The beams traveled inside a 6.3 km tunnel below the suburbs of Hamburg from 1992 until 2007. The high-energy protons were forced on their circular journey by superconducting magnets (Chap. 9).

16. David Gross (b. 1941), Frank Wilczek (b. 1951), and David Politzer (b. 1949) received the Nobel Prize in Physics in 2004 "for the discovery of asymptotic freedom in the theory of the strong interaction."

17. A readable account given by one of the creators can be found in: Frank Wilczek, *Asymptotic Freedom, from Paradox to Paradigm,* Proceedings of the National Academy of Sciences of the United States of America, June 2005, page 8403.

18. One implication of Heisenberg's uncertainty relation is that one can never determine a change of energy ΔE within a time interval Δt with infinite precision. The product $\Delta E \times \Delta t$ is always greater than $h/4\pi$, where $h = 6.626 \times 10^{-34}$ Js is Planck's constant. This allows a particle-antiparticle pair to spontaneously pop

into existence, but it forces the pair to recombine and disappear within the time span given by Heisenberg's uncertainty principle. Note, however, that these virtual processes are not completely anarchic; a number of conservation laws, among them that for charge, must be satisfied. That's, for example, the reason this works for a particle and its antiparticle.

8

Spearheading Charm

Shortly after Veksler and McMillan discovered the principle of phase focusing (Chap. 4) and before it was used to convert cyclotrons to synchrocyclotrons, small electron synchrotrons were built in several places.

Electron Synchrotrons

Already in 1946, Frank Goward and his colleagues boosted the energy of their small 15 cm betatron from 4 MeV to 8 MeV by adding a radio-frequency system that "pulled" the electrons along while increasing the magnetic field. A few months later, at General Electric in Schenectady, a 70 MeV electron synchrotron with a diameter of 60 cm was taken into operation. Luckily its vacuum chamber was made of glass so that the accelerator operators could see light coming from the beam, the first observation of *synchrotron radiation*.[1] Also McMillan in Berkeley, one of the inventors of phase focusing, constructed a 330 MeV electron synchrotron, which reached its designs specification early in 1949.[2] It was used to produce X-rays and pions thanks to its energy being above the threshold for the creation of a π^+-π^- pair. From this point on electron synchrotrons were proliferating world-wide. Even the magnets from the quarter-scale model to ensure the feasibility of the Bevatron, were shipped to the California Institute of Technology near Los Angeles, where they reincarnated as a 1 GeV electron synchrotron.

All these synchrotrons were built before Livingston and colleagues came up with the idea of alternating-gradient focusing (Chap. 6). Even though Livingston had proton synchrotrons in mind, the first alternating-gradient synchrotron accelerated electrons rather than protons. Robert Wilson[3] used the idea for a 1 GeV electron synchrotron at Cornell University in the US. This machine was the precursor of many electron synchrotrons since, among them the 6 GeV Cambridge Electron Accelerator (CEA) near Boston and the 7 GeV Deutsches Elektronen Synchrotron (DESY) in Hamburg.

The emission of synchrotron radiation from accelerated electrons, first observed in Schenectady, had a good and a bad side to it. The bad news first: the electrons lose energy that the radio-frequency system has to replenish. This does not play an important role in low-energy machines, but becomes the dominant factor at very high energies. The good news is that the emission reduces synchrotron and betatron oscillations (see Chap. 4) and that causes the beam sizes to shrink. On the other hand, the emission of radiation is a quantum mechanical process and therefore random, which slightly excites these oscillations. The balance of excitation and damping determines the beam sizes in all electron synchrotrons. This makes operating electron rings rather robust, because disturbances quickly damp away. The equilibrium beam size and how quickly the equilibrium values are reached are normally designed into the accelerator by judiciously choosing parameters of the magnet and the radio-frequency system.

So far, all accelerator-based experiments used a high-energetic beam to bombard a stationary target. Such a *fixed-target* collision is illustrated in the foreground on Fig. 8.1. It has the advantage that the target can be made rather dense, such that one can expect many collisions. Moreover, all collision products come out in a well-defined direction. On the other hand, much of the energy invested in accelerating the beam is "wasted", because the total momentum in the collision must be preserved and the collision products therefore move with high speed in the direction of the incident beam. And there is a lot of energy needed to maintain this motion. Would it not be nice to use all the energy in the beam to create new particles—kaons, pions, and the like?

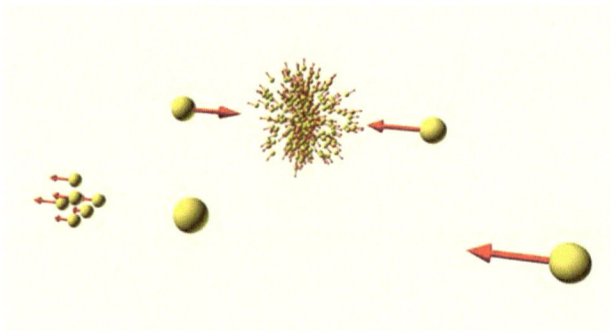

Fig. 8.1 Fixed-target (front) and head-on collision (back)

Great Idea: Storage-Ring Collider

Now imagine that we arrange head-on collisions of two beams, as shown further back on Fig. 8.1. If the initial momenta of the two beams are equal but have opposite sign, their sum is zero, and no energy is needed to maintain some average motion after the collision. All energy, available from the beams, is available to create new particles. For collisions at very high energies, the gain can be several hundred times compared to corresponding fixed-target collisions. But the promise to reach very high energies comes at the expense of much reduced collision rates, because the density of particles in beams is many times lower than in solid targets. In order to compensate this disadvantage, the beams must collide very frequently. This is accomplished by first accelerating the beams to their collision energy and then storing them for long times at this energy in a so-called *storage ring*. Moreover, we need one ring for each of the colliding beams.

Ideas regarding colliders appeared in print in the early 1950s before Gerard O'Neill[4] from Princeton convinced his colleagues at Stanford that probing the substructure of electrons is a worthwhile enterprise. After all, atoms and nuclei showed a substructure, so maybe electrons do so, too. He tried to find that out by scattering counter-propagating electrons off of each other in the two storage rings shown in Fig. 8.2. Each ring has a circumference of 12 m that tangentially touches the other ring in a shared section where the two beams collide head on. A camera helped to ensure that the beams actually collided. Two acceleration systems replenished the energy lost by emitting synchrotron radiation. He filled the rings with 500 MeV electrons from the Mark III linac (Chap. 7) that was already up and running at the time. Synchrotron-radiation damping,

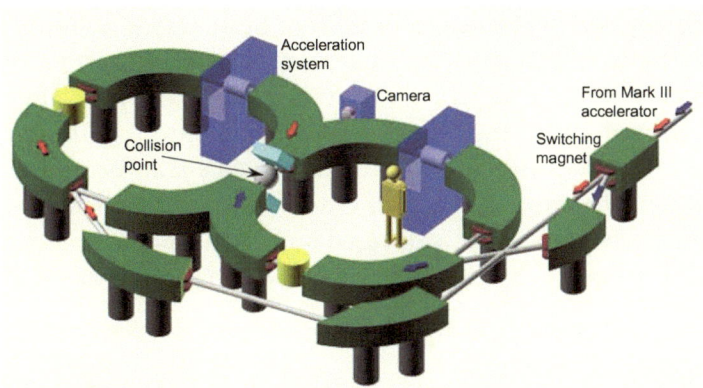

Fig. 8.2   The Stanford electron-electron collider

supported by very fast-pulsed magnets, ensured that the beams could be stored in the rings with millimeter-sized beams. Unlike in synchrotrons, beams in storage rings often circulate for hours and therefore special attention had to be paid to achieve good vacuum conditions to prevent the beam from crashing into gas atoms flying around inside the beam pipe. The region, where the two beams collide, was surrounded by detectors that recorded whatever came out from the collisions.

The construction went well, but being the first machine of its kind, many new phenomena made themselves known, such as instabilities due to high beam intensities. It took a while to understand and control these novel effects but by 1962 experiments started for good. A typical cycle of operating the collider was as follows: electrons were accelerated in the Mark III linac and first injected into one ring and then into the other. While the experiments went on, the linac could be used for other experiments at the end of the linac, before it was needed for the next fill, typically 30 min later. The experiments in the rings showed that if electrons have a size at all, it must be smaller than one tenth of the diameter of a proton. That's pretty point-like. Remarkably, in all experiments ever since the electron always appeared point-like.

O'Neill's collider needs two independently operated rings, one for each of the counter-propagating electron beams. At about the same time, on the other side of the Atlantic, Bruno Touschek in Frascati near Rome realized that he could store counter-propagating electrons and their antiparticles—positrons— in the same magnet system. This works because magnetic forces depend on the product of particle velocity and charge. If both velocity and charge have opposite signs, the particles follow exactly the same trajectory. Hence he could store both beams in the same ring. To prove his idea, he built a small ring with a circumference of four meters, called AdA ("anello di accumulazione", Italian for accumulator ring). By 1961 he managed to store both beams simultaneously in the little ring and made them collide.[5] The beam intensity, however, was too low to do useful experiments, but it served its purpose as proof-of-principle protoype so well that a larger collider, called ADONE was planned. With a circumference of 105 m it stored electrons and positrons with energies of up to 1.5 GeV each. As a novelty, it featured a *separate function* magnet structure consisting of different magnet types. One type of magnet deflected the beam and a second type, quadrupoles, focused it. This made the magnets simpler to build and less expensive compared to magnets that do both simultaneously. ADONE started operations in 1969.

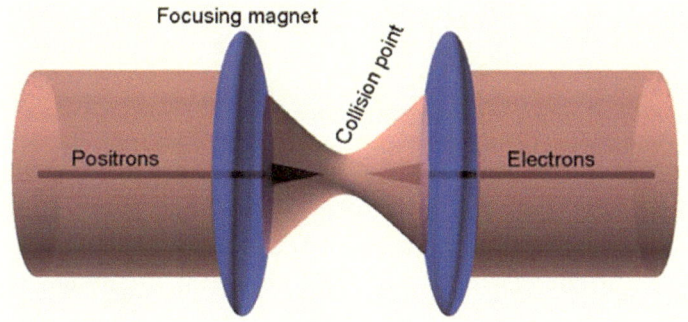

Focusing magnet

Collision point

Positrons

Electrons

Fig. 8.3 Low-beta insertion

Great Idea: Small Beams at the Collision Point

The Cambridge Electron Accelerator, often referred to as CEA, was conceived and taken into operation as a synchrotron in 1961. Sixty times per second it accelerated 10^{11} electrons to an energy of 6 GeV. By 1966 Livingston had relocated from Brookhaven to Cambridge near Boston and, together with his colleagues, had figured out a way to simultaneously store electrons and positrons in the ring. But under normal conditions, the rate of collisions was rather low, because the beams were too large. In order to remedy the problem Gustav Voss[6] came up with the idea of the *low-beta insertion*,[7] where several focusing magnets, in Fig. 8.3 illustrated as lenses, squeeze the beams to very small sizes just before they collide. This works much like a telescope demagnifying an optical image. Only, in an accelerator magnets take the role of lenses to squeeze the beam sizes before the beams are made to collide. With small sizes the chance of an electron actually colliding with a positron is greatly increased. Practically all colliders use this concept ever since. In Cambridge, however, it was not possible to permanently operate with the low-beta insertion, which is why it was placed in a short beam line, called the bypass, that was installed parallel to a section of the ring. Only when the beams were ready for collisions, were they directed through the bypass, an operation that was extremely difficult to perform. Nevertheless, the low-beta insertion became a lasting contribution to accelerator science.

Having earlier worked on O'Neill's collider at Stanford throughout the 1960s, Burton Richter by then having moved to SLAC, unsuccessfully tried to secure funding for an electron-positron collider. Only in 1970, aided by a helpful administrator, he found a loophole in the funding regulations that made it possible to build his collider without special funding, provided he declared the whole machine an "experiment." This trick did not create new money, but made it possible to redirect some of the existing funds to building a ring, though he had to strip down its design to a bare minimum. Here's what he came up with.[8]

SPEAR

The Stanford Positron Electron Asymmetric Ring (SPEAR) was a 234 m oval that operated at energies up to 4.5 GeV per beam. Its design incorporated practically every feature developed in the earlier rings. Semi-circular arcs with separate-function magnets connect low-beta insertions to ensure a high collision rate for detectors. One electron and one positron packet, supplied from the SLAC linac, circulated in SPEAR and collided inside the low-beta insertions. A host of automatic control and feedback systems made sure that the beams circulate stably for hours, while the detectors collected data.[9]

One of the collision points was surrounded by the Mark I (folks in Stanford had a penchant to call versions of large technical installations "Mark") detector shown in Fig. 8.4. It covered practically every direction that collision products could escape to. As the particles move radially away from the interaction point in the center of the beam pipe, a trigger counter notices their presence and a spark chamber provides information about their trajectories. A second layer of scintillator triggers follows just inside the coil of the magnet. Outside the coil lead scintillators provide information about the energy of escaping electrons. Only muons can penetrate the iron yoke and are detected in the spark chambers on the outside.

Figure 8.5 illustrates[10] how new particles are created from the incident electrons and positrons in the first place. An electron, indicated by a forward-

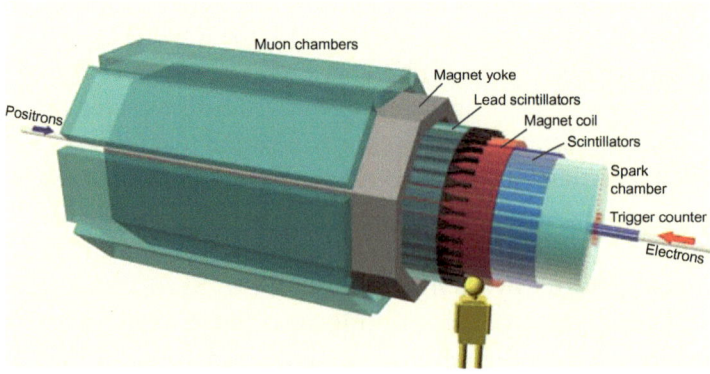

Fig. 8.4 Sketch of the Mark I detector

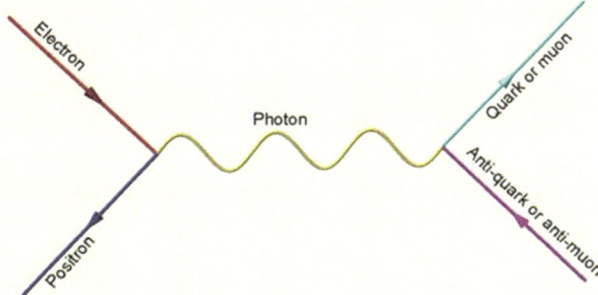

Fig. 8.5 Electron-positron production reaction

pointing arrow head, comes from the upper left. Likewise, a positron, indicated by a backward-pointing arrow head, comes from the lower left. At their meeting point electron and positron annihilate each other and give rise to a photon, indicated by the wavy line. This photon represents the fireball of pure energy that spontaneously gives birth to a particle-antiparticle pair in the vertex on the right-hand side in which muon pairs or quark-antiquark pairs can appear. As a matter of fact, the count rate of creating these pairs is determined by the electric charge of the produced particles. And the count rates to create different quarks just add up. Thus the the ratio of the count rate of finding anything that's made of quarks—hadrons—and the count rate of muons depends on the number of quarks and their electric charges. Note how this provides a method to count the number of quarks.

Earlier measurements in ADONE and at the CEA bypass already indicated that just the three quarks u, d, and s are not enough to explain the observed count rates. The experimental rates were three times higher than those predicted by the three-quark model. This was a clear indication that each quark "counts as three" and the three colors in QCD were a hot candidate to account for it. What was surprising from the earlier measurements was that the count rates from quark-antiquark pairs were growing with increasing beam energies. Richter and his colleagues at SPEAR went ahead to figure out what was going on.

November Revolution

Richter and his collaborators at SPEAR started by scanning the energy of the beams from 1200 MeV to 2400 MeV in steps of 100 MeV and diligently recorded the count rates for hadrons and for muons at each step. The data did not seem to indicate drastic differences to the results from ADONE or CEA, only a small surplus of hadrons appeared at 1500 MeV and 1600 MeV. This surplus was not very big, but in November 1974 the collaboration agreed to take a closer look by taking ten times smaller energy steps in that region. Near 1550 MeV they found that the count rate went up tenfold and more. Zooming into the region with even smaller step sizes they found the count rate going up thousand-fold. That was remarkable in itself, but at the same time stepping the beam energy just a few MeV to either side brought the count rate down to the base level. This was some narrow resonance! The small width of the resonance indicated—by quantum-mechanical reasoning—that its lifetime is very long. If the lifetime is that long, it's a particle. Moreover, its mass is given by the sum of the beam energies, which makes it close to 3100 MeV. The collaborators from SPEAR named it ψ (psi). Practically at the same time Samuel Ting's group, working on the AGS in Brookhaven, found the same particle, only they called it J, by reconstructing electron-positron pairs created by smashing protons into a beryllium target.[11] Both groups rapidly wrote up their findings and the articles appeared back-to-back a few weeks later. Ever since the new particle was known as J/ψ.

A hint at what this resonance is came from scanning the beam energy in SPEAR over a wider range. Lo and behold, soon another resonance, named ψ' with a mass of 3700 MeV and almost as narrow as the ψ was found. Adjusting the beam energies to half the resonance's mass, the sum added up to its mass and many ψ' could be produced. Subsequently analyzing their decay led to the hypothesis that the ψ is made up of a quark and an antiquark that revolve around each other, much like an electron revolves around an atomic nucleus. Only here the particles are quarks and the force is the strong nuclear force. The question, however, remains: which quarks are involved? Since all known hadrons were already accounted for by the u, d, and s quark, this must be an additional fourth quark.

Already a few years earlier, Sheldon Glashow, John Iliopoulos, and Luciano Maiani suggested the existence of such a fourth quark—they called it *charm*—to explain the scarcity of some kaon decays. After determining more and more properties of the J/ψ and the ψ' evidence was mounting that the J/ψ resonance is indeed a bound state of a charm and an anti-charm quark. This resonance, collectively with its excited states, were given the name *charmonium*.

A charmonium resonance lives quite long—the resonance is very narrow—but not forever and when it does decay, the charm quark and the anti-charm quark annihilate and often turn into new quark-antiquark pairs that "somehow" turn into a spray of hadrons coming out back-to back in so-called *jets*. First indications of jets were actually seen in SPEAR, though they are much more pronounced at higher-energy accelerators. Still, these first indications added to the evidence that there was a new quark in town.[12]

But what prevents the quarks from appearing alone[13] and how does the "somehow" work and turns quarks into hadrons and jets?

Quarks to Hadrons to Jets

Born in the process from Fig. 8.5 and illustrated by the insert in the top right in Fig. 8.6, two quarks rapidly move apart. During the moment of birth they are asymptotically free, as discussed towards the end of Chap. 7; no force lines tie the quarks to each other. But as they move further apart, the strong nuclear force, represented by the thick green lines, increases. Energy from the quarks is used to build up the field lines. At some point, the intensity of the field becomes so large that a new quark-antiquark pair spontaneously appears; another quantum-mechanical trick of nature. Each of these two new quarks joins up with one of the original quarks and forms a meson, illustrated by two dumbbells towards the right in the insert in Fig. 8.6. They are held together by the color force. Since the quarks inside of the meson do not necessarily move in the same direction, they start to separate. This causes the field between them to increase and give rise to another quark-antiquark pair. This process repeats until all of the initial energy is converted into mesons and other hadrons. Since all newly created particles inherit the momentum of the initial quarks, they show up in preferred directions as jets.

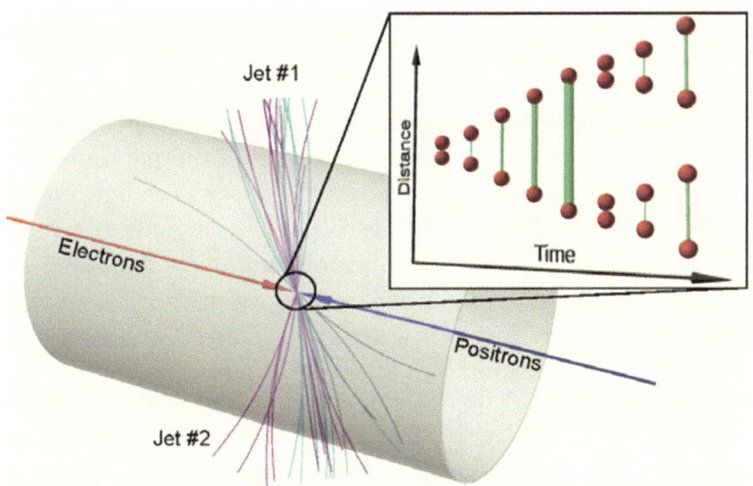

Fig. 8.6 Jets and hadronization

Finding the charm quark and seeing them forming jets was a spectacular success. It also made the theoreticians rather happy, because now the strange quark had a partner in the charm quark, much like the up and down quarks form a partnership. Thinking of it, even an electron has a neutrino as partner. Likewise the muon has its neutrino as partner. Considering the charge of these particles hints at even deeper connections. Let's therefore assemble these particles—all of them having spin 1/2 and are therefore fermions—in a table ordered horizontally by their charge and vertically such that we place the particles discovered earlier in the first row and those discovered later in the second row. We recover the top two shelves with particles from the figure on the inside of the front cover.

Charge	2/3	−1/3	−1	0
1. generation	up quark	down quark	electron	electron neutrino
2. generation	charm quark	strange quark	muon	muon neutrino

Remarkably, a relation among these particles appears: two quarks with fractional charges 2/3 and −1/3 and two leptons with charges −1 and 0 belong to the same *generation of fermions*.[14]

This appealing classification of the subatomic world did not, however, last very long.

Tau Lepton

Martin Perl,[15] also a member of Richter's collaboration at SPEAR, had set his mind on figuring out the difference between the two known types of leptons: electrons and muons. However, this did not work out as expected. Instead, he tried to find out whether there might be an additional lepton. Since leptons only decay into lighter leptons, he surmised that a unique signature of their existence were one electron and one muon with opposite charges escaping from a collision in SPEAR.[16] Jointly with his colleagues, carefully grinding through the events, he actually found a number of candidates that were eventually validated as originating from a new lepton that was named *tau*. It behaves just like electrons and muons, but with a mass of $1.77\,\text{GeV}/c^2$ it is much heavier.

But how do we fit in the tau into our table of fermions; there is no free slot in the table. It looks like we have to open another row in it to accommodate the tau. But that immediately raises the question of how to fill the remaining three slots for the tau-neutrino and two additional quarks.[17] The story of the discovery of the fifth quark,[18] called *bottom*, belongs to the next chapter, but once its mass was known, several electron-positron colliders, modeled after SPEAR, stood ready to explore it.

Larger Electron-Positron Collider

Spurred by the success of SPEAR, the German research laboratory DESY in Hamburg, Germany decided to build the storage-ring collider DORIS and feed it with beams from the venerable DESY synchrotron. DORIS successfully explored the J/ψ and its excited states. Unfortunately its maximum beam energy of 4.45 GeV was insufficient to explore systems where the fifth quark is involved. After an upgrade, however, DORIS played an important role in exploring the bound state of a bottom quark and a bottom antiquark that was called upsilon (Υ). The experimenters also analyzed the lowest excited state, called Υ', much like SPEAR explored the J/ψ and its lowest excited state, the ψ'.

After the discovery of the bottom quark, Cornell University was authorized to build the Cornell electron storage ring (CESR) that was filled from the 10 GeV synchrotron. CESR's maximum energy of 12 GeV was sufficient to explore many more excited states of the upsilon and even produce mesons composed of the fifth quark and one of the first four quarks. Tongue in cheek these mesons were said to expose "bare bottom."

After the fifth quark was discovered, the community expected to find the sixth quark, which had already been named top-quark, in the next generation electron-positron colliders. Three machines reaching beam energies between 15 and 30 GeV became operational towards the end of the 1970s: PETRA in Hamburg, PEP at SLAC, and TRISTAN in Japan. Neither found the top quark, whose mass turned out to be outside their reach. PETRA, on the other hand, found something else.

Gluons

Already SPEAR found indications for jets; hadrons rushing away from the collision point seemed to remember the momentum of the original quark and the antiquark and moved in opposite directions. Due to the low beam energies available in SPEAR the jets were not very pronounced. In PETRA, at much higher energies, the jets became much more distinct. Even more spectacularly, once in a while, three jets showed up in the reconstructed events.[19] What could be responsible for the third jet? After all there are only two quarks around at the moment of birth. This came not as a surprise to John Ellis and his colleagues,

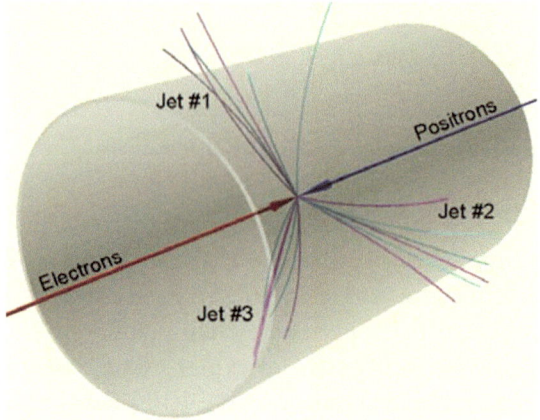

Fig. 8.7 A three-jet event

who previously had theoretically analyzed processes where one quark of the original pair emits a gluon and thereby changes its color. The gluon then creates a quark-antiquark pair, which subsequently turns into the hadrons of a third jet. Such a three-jet event is illustrated in Fig. 8.7. This first observation of gluons was very strong evidence that the theoreticians' analysis, which was solidly based on QCD, was correct and that therefore QCD is the right theory to describe the strong nuclear force. Confidence in the theoretical framework was growing.

We've already mentioned the discovery of the fifth—bottom—quark. It's about time to catch up with developments in large proton synchrotrons, which is where it happened.

Notes

1. For a readable account by one of those who did it: John Blewett, *Synchrotron Radiation—Early History,* Journal of Synchrotron Radiation 5, (1998) 135. We will come back to this topic in Chap. 13.
2. Edwin McMillan, *The History of the Synchrotron,* Physics Today, February 1984, page 31.
3. After leading the construction of the electron synchrotron in Cornell, Robert Wilson (1914–2000) became the first director of Fermilab and led the construction of its main ring (Chap. 9).
4. Gerard O'Neill (1927–1992) tells the story himself in: Gerard O'Neill, *Particle Storage Rings,* Scientific American, November 1966, page 107.

5. The problem of getting electrons and positrons into the ring was solved by placing a thin tantalum target into the ring and bombarding it with electrons to generate very high-energy photons that, in turn, create electron-positron pairs. Keeping his fingers crossed, Touscheck and colleagues hoped that some of those could be stored. Remarkably, this somehow worked but the intensity of stored beam was very low. For the full story of AdA, see: Carlo Bernardini, *AdA: the First Electron Positron Collider,* Physics in Perspective 6 (2004) 156.

6. After working on the Cambridge electron accelerator, Gustav Voss (1929–2013) moved to DESY and led the construction of PETRA (in this chapter) and HERA.

7. Here *beta* refers to the beta-function, a somewhat technical concept that accelerator builders use to describe the beam size, among other things.

8. Shawna Williams, *The Ring on the Parking Lot,* CERN Courier, May 2003, https://cerncourier.com/a/the-ring-on-the-parking-lot.

9. Alan Litke and Richard Wilson, *Electron-Positron Collisions,* Scientific American, October 1973, page 104.

10. The graphical representation of interactions among elementary particles shown in Fig. 8.5 is commonly referred to as *Feynman diagram.*

11. Here "at the same time" was really touch and go: Ting's way of finding electron-positron pairs in the debris of protons smashing into a target was much more difficult than counting events in SPEAR. Therefore Ting's team in Brookhaven meticulously checked and rechecked their results. While the rumors that Richter was on to something grew in intensity, Ting was actually on a plane to California to participate in a meeting at SLAC. His team, however, managed to pass this information on to him such that Ting received it at the airport in San Francisco. Over night, they prepared plots to show their data and Ting wrote up a report the next day while still being a SLAC. He went back to Brookhaven the same evening and hand-delivered it to the editor of a journal, where it arrived at practically the same time as Richter's. Both reports were published back to back in the same journal and the rest is history. Burton Richter (1931–2018) and Samuel Ting (b. 1936) shared the Nobel Prize in Physics in 1976 "for their pioneering work in the discovery of a heavy elementary particle of a new kind."

12. Roy Schwitters, *Fundamental Particles with Charm,* Scientific American, October 1977, page 56.

13. Yoichiro Nambu, *The Confinement of Quarks,* Scientific American, November 1976, page 48. Kenneth Johnson, *The Bag Model of Quark Confinement,* Scientific American, November 1979, page 112.

14. Sheldon Glashow, *Quarks with Color and Flavor,* Scientific American, October 1975, page 38.

15. Martin Perl (1927–2014) received the Nobel Prize in Physics in 1995 "for the discovery of the tau lepton" and "for pioneering experimental contributions to lepton physics." For a first-hand account see: Martin Perl and William Kirk, *Heavy Leptons,* Scientific American, March 1978, page 50.

16. A tau and an anti-tau also decay into an electron-positron pair or a muon-anti-muon pair, but that can also happen in other reactions where no taus are involved. On the other hand, an electron together with an anti-muon can only come from a tau-anti-tau pair.

17. David Cline, Alfred Mann, and Carlo Rubbia, *The Search for New Families of Elementary Particles,* Scientific American, January 1976, page 44.

18. The upsilon particle with a mass of approximately $9.5\,\mathrm{GeV}/c^2$, consisting of a bottom quark and an anti-bottom quark, was discovered at Fermilab in 1977.

19. I. Flegel, P. Söding, *Twenty-five Years of Gluons,* CERN Courier, November 2004, Available from https://cerncourier.com/a/twenty-five-years-of-gluons/.

9

The Tevatron and Generation Matters

Following the discovery of the Omega-minus in 1964 and the success of the quark model, the Atomic Energy Commission in the US issued the recommendation to build a proton synchrotron with an energy of at least 200 GeV. After a nation-wide site contest, the National Accelerator Laboratory (NAL)[1] was located on the outskirts of Chicago and Robert Wilson from Cornell was appointed its founding director in 1967. The budget for the accelerator was extremely tight, such that Wilson had to find ways[2] to pack the large accelerator into the limited funds. One way is to make the magnets smaller and simpler.

Great Idea: Cascaded Accelerators

The maximum field in conventional magnets is limited by the material properties of iron. Therefore reaching higher energies requires more magnets and that makes the rings larger.[3] At the same time particles must also reliably circulate at the, usually much lower, injection energy, where the magnetic field is low. Maintaining good field quality both at low injection energy and at maximum energy becomes very difficult. One way out of this dilemma was earlier suggested by Matthew Sands from the California Institute of Technology near Los Angeles. He proposed to use a sequence of rings of increasing size, shown in Fig. 9.1, each with magnets operating in a region where good field quality can be easily maintained. He called this scheme *cascaded accelerators*. It has an additional benefit, because the beam size at high energies is smaller than at low energies. This makes it possible to use smaller magnets in the higher-energy rings, and smaller magnets are more economical to build and to operate.

© The Author(s), under exclusive license to Springer Nature Switzerland AG 2024
V. Ziemann, *Beams*, Copernicus Books,
https://doi.org/10.1007/978-3-031-51852-2_9

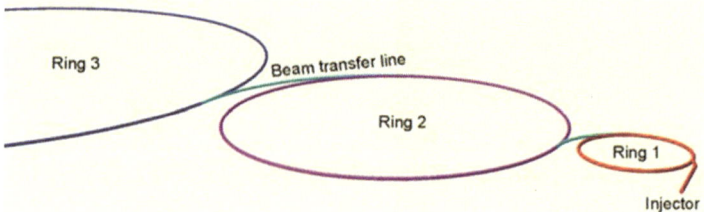

Fig. 9.1 Cascaded accelerators

Fermilab

Wilson used the cascading scheme to feed protons to the large ring. The beams are born in an ion source, shown near the top of Fig. 9.2, and accelerated to 0.75 MeV in a Cockroft-Walton accelerator. Then they are passed to a linear accelerator of the type Alvarez had come up with for the Bevatron (Chap. 5). The linac accelerates the protons to an energy of 200 MeV before they are injected into the booster synchrotron. It has a circumference of 474 m and accelerates the protons to 8 GeV. Thirteen consecutive proton batches from the booster are then injected into the *main ring* to fill its circumference of 6.3 km.

Instead of using magnets that deflect and focus the beam simultaneously, Wilson used dedicated magnets to deflect the beam and others, quadrupoles, to focus the beam. For obvious reasons this type of magnet assembly was referred to as a *separate function lattice*, where lattice is jargon for a sequence of magnets. In this way the magnets became simpler to design and assemble and this made them less expensive. Moreover, since they only had to perform one task—either deflecting or focusing—their performance was easier to optimize. The magnets became smaller and could be powered from smaller power supplies, which made the motor-flywheel-generator combo used in earlier accelerators obsolete.

After three years of construction, the main ring was ready to receive the first beam in 1972. After some hiccups during commissioning, mostly owing to the frugal design of the machine, protons eventually reached the design energy of 200 GeV. What is even more impressive, they could operate at twice their original specifications and protons were accelerated to 400 GeV about half a year later. Eventually protons even reached 500 GeV, though this required prohibitive levels of electric power and that was impossible to finance during the oil crisis of the mid-1970s.

Several times per minute a new beam of protons was available at the peak energy, ready to be extracted from the main ring and delivered to one of three

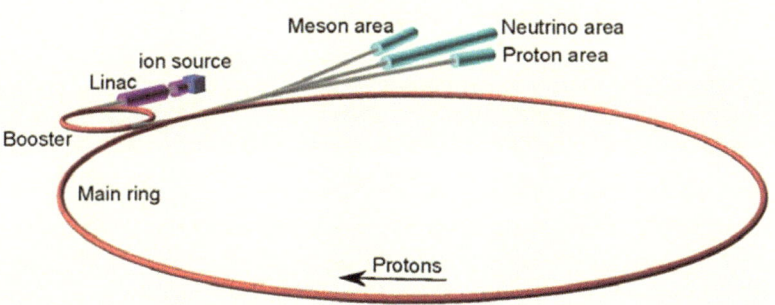

Fig. 9.2   Fermilab

external experimental areas, shown on the upper right in Fig. 9.2, where it impinged on a variety of targets. There it generated secondary beams that were further analyzed in bubble chambers or special magnets that could select specific particles, which were subsequently counted in spark chambers or scintillators.

Upsilon and b-Quarks

Over the time span of several years one of the experiments, labeled E288 and led by Leon Lederman, had assembled magnets, counters, and other detectors. Their experiment allowed them to identify individual muon pairs—one positive and one negative muon—coming from their target. By 1977 they found that the sum of the muon energies frequently added up to 9.5 GeV, which they interpreted as a new particle they gave the name upsilon.[4] Once the energy of the upsilon was known, the electron rings DORIS and CESR tuned their beam energies to the resonance and joined the work to determine its properties, such as its spin and the charge of its constituents. This confirmed that the upsilon was indeed a quark-antiquark state of the fifth quark, the bottom, or b-quark that we already mentioned in Chap. 8.

At this point, two members of the third generation of particles were known: the tau lepton and the bottom quark, which turned out to carry electric charge $-1/3$. What's missing is the tau neutrino and a sixth quark with charge $2/3$. The larger electron rings PETRA, TRISTAN, and PEP from Chap. 8 did not find the latter despite reaching high energies. The sixth quark, provisionally called *top* quark, must indeed be very heavy. Luckily the Fermilab management had already engaged in developing means to double the energy of their accelerators once again.

Great Idea: Superconducting Magnets

The magnets of the main ring already operated at the maximum field possible with iron-based magnets. Since the higher-energy accelerator had to fit into the existing tunnel, it had to use a different magnet technology to reach higher fields. Superconducting magnets turned out to be the solution.[5]

Around 1908 Heike Kamerlingh Onnes[6] in Leiden in the Netherlands wondered how different materials behaved at very low temperatures and needed something, preferably a liquid, to cool them down. Practically every cooling agent he tried became solid at very low temperatures. Only helium, which is a gas at ambient temperatures, turned liquid a few degrees above absolute zero temperature,[7] but never became solid. Helium thus turned out to be the perfect cooling agent for his explorations of material properties, one of which was the electrical resistance of metals. Most metals showed a continuous decrease of the resistance as he lowered the temperature, but in a wire of mercury the resistance abruptly dropped to zero and stayed there as he lowered the temperature further. Onnes called this phenomenon *superconductivity*.

This abrupt loss of resistance implies that electric currents, even very large ones, flow in the wire unimpeded and do not heat up the wire. Ideally, a current launched in a superconducting wire will flow forever, even with the power supply disconnected. Therefore, using superconducting wires in a magnet only requires energy to cool the wires, but no energy to maintain the current flow. This makes such magnets rather economical. Contrast this to a magnet using normalconducting wires. It requires a continuously operating power supply to keep the current flowing, because a large amount of energy heats the wire, which is what made operating the main ring at 500 GeV so expensive.

Energy Doubler

The potential benefits of superconducting magnets were well-known to Wilson and his collaborators and, as the main ring was being installed, they initiated a research program to develop superconducting magnets that would take the beam to 1000 GeV.

This program had to address a number of rather intricate technical questions, such as what type of wires to use. The wires had to work both in the superconducting state, but also when the wire becomes normalconducting after getting too warm. In such a transition, called a *quench*, the superconductor suddenly loses its ability to carry currents without resistance. Yet, even then it has to carry the same large currents in without burning up. Moreover, the large currents, needed to excite the magnetic fields, exert large forces on the superconductors which must not move, because the shape of the coils determines the magnetic field quality. These questions were solved by constructing more than one hundred short prototype magnets and several longer versions. This took several years, but eventually the magnets were deemed ready. More than a thousand superconducting magnets were installed below those of the main ring and cooled by one of the world's largest refrigeration plants for liquid helium. This plant was erected on the Fermilab site.

This superconducting ring in the same tunnel as the main ring was called the *energy doubler* or the energy saver, because the electrical power needed to operate it was only a fraction of what was used to operate the main ring at top energy. It was ready for action by 1983. In its normal mode of operation the main ring was filled from the linear accelerator and the booster before the beam energy was increased to only 120 GeV to save electricity in the magnets of the main ring. Since both rings have the same circumference, at this point the entire beam was transferred to the superconducting ring, which subsequently increased the energy to 800 GeV and later even to 900 GeV. At this energy the beam was extracted and guided to the experimental areas. Since the energies are close to 1000 GeV, which is also referred to as 1 TeV, the accelerator was called *Tevatron*.[8]

The new ring allowed experiments at unprecedented energies in the experimental areas where the beam crashed into targets. But since only the beam is moving and the targets are stationary, the energy to create new particles is limited. Just as for electrons, colliding two beams head on would make all energy available. And for protons that technology was developed at CERN where the *Intersecting Storage Ring* (ISR) was constructed during the late 1960s.

Intersecting Storage Ring

Already during the final stages of assembling the CERN PS (Chap. 5) discussions regarding the next accelerator started. Instead of constructing a large synchrotron, similar to Fermilab's, the CERN management decided to build a proton-proton collider, the *Intersecting Storage Ring*, or ISR. It is conceptually similar to O'Neill's electron rings in Stanford (Chap. 8), but for colliding protons. Smashing accelerated proton beams into each other head-on promised unprecedented collision energies from which new particles would emerge.

Under the leadership of Kjell Johnsen construction of the ISR started by the end of 1966 and first beams were stored four years later. The accelerator, shown in Fig. 9.3, consists of two rings with a circumference of 942 m each into which protons from the PS with energies of up to 26 GeV were injected. In order to accumulate very high beam currents, the old injector linac was replaced

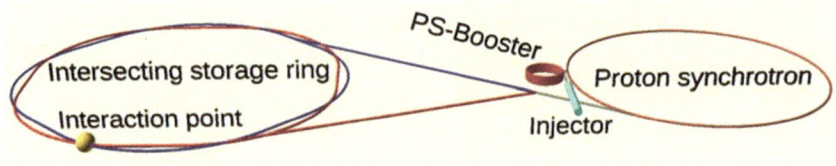

Fig. 9.3 Intersecting storage ring

by a new linear accelerator and four booster synchrotrons, mounted on top of each other. This four-fold synchrotron, called the *PS booster*, made it possible to increase the beam intensity in the PS ten-fold. Moreover, an ingenious scheme, called *stacking*, was invented. It allowed Johnsen and his crew to pile more and more protons into the ISR by employing the radio-frequency system to move a freshly injected beam a little bit sideways inside the ISR such that the next pulse from the PS could be injected into the just liberated space. Several hundred injections could be accumulated to store record numbers of 10^{15} protons in each of the rings. This was absolutely necessary in order to reach competitive count rates in the experiments, because beams are less dense than solid targets. The beams collided in eight interaction points where the two rings cross each other. Adjacent superconducting focusing magnets.[9] reduced the beam size and thereby increased the collision rate further. Since filling the two rings took the better part of a day, the time during which collisions took place had to be long, up to two full days. This was made possible by an exceptionally well-designed vacuum system that minimized collisions with stray gas molecules. Their density was comparable to that in the outer atmosphere of the earth.

As the first proton-proton collider ever, the ISR exhibited a number of unforseen phenomena, especially related to the very large beam intensities. At one point the beam created a feedback loop with itself which became unstable, much like the feedback whistle at a rock concert when a microphone gets too close to a loudspeaker. In the ISR, special magnets and additional radio-frequency components were developed to alleviate these problems. Other difficulties arose, because the two beams perturbed each other if they collided with some sideways displacement. Novel methods to center the beams at the collision point solved this problem. After the time the ISR was operating stably, specially developed sensors detected fluctuations in the beam current, so-called *Schottky noise*, which is due to the finite number of protons in the beam. These and many other techniques developed at the ISR proved essential for all subsequently built accelerators.[10]

The experiments for the ISR were planned in the 1960s, before the advent of deep-inelastic scattering results from SLAC. At the time, practically everybody expected that the reaction products would leave the collision point very close to the beam axis, just like all debris in fixed-target experiments moves mostly in the same direction as the incident beam. Therefore many smaller experiments were assembled around the beam pipe to do what was called "keyhole physics." Then, once the beams collided, to almost everybody's surprise, many particles, especially pions, escaped from collision point almost perpendicular to the direction of the beams. Soon this observation was related to the deep-inelastic-scattering experiments with electrons at SLAC (Chap. 7) which had previously found point-like quarks inside the protons. Apparently at the ISR, instead of electrons, quarks inside the protons of one beam scattered off of quarks inside the protons of the other beam. To analyze the collision products flying off in all directions a large detector that surrounds a collision point was needed.[11] Unfortunately for the ISR, by the time this was realized, funds for a new detector were unavailable, because plans had matured to build a 300 GeV synchrotron, the *Super Proton Synchrotron* (SPS).

Super Proton Synchrotron

It was decided to construct the SPS close to the CERN site rather than creating a new laboratory in one of the CERN member states. This became possible through advances in tunnel-boring technology during the 1960s, because the accelerator could be located in a 40 m deep and 6.9 km long tunnel underneath the outskirts of Geneva. The accelerator was serviced by six vertical access shafts connecting the tunnel to the surface. Like the main ring at Fermilab, it used separate magnets to focus and to deflect the beam, which made it possible to increase the achievable energy from 300 GeV to 400 GeV by the time the SPS was commissioned in 1976. Later even 450 GeV beams became available. Controlling the large number of components was made possible by one of the first *local area networks* of distributed computers. The operators in the control room accessed the control system through early versions of *touch panels*,[12] the precursors of those found in our smartphones.

An operation cycle of the SPS begins by injecting multiple pulses of 26 GeV protons from the PS, which were stored one after the other in the SPS. Once most of its circumference was filled, the energy was increased up to the desired maximum when the beam was extracted and guided to two experimental areas, each with a number of beam lines and targets. The very flexible control system made it possible to either extract the entire beam in a single pulse or to slowly spill it over several seconds, depending on the needs of the experiments. For a number of years, the SPS operated in fixed-target mode and produced a variety of secondary beams, including kaons and magnetic-horn-focused (Chap. 6) neutrinos, which were subsequently directed to bubble chambers and other counter-based experiments.

Towards the end of the 1970s Carlo Rubbia[13] and others started dreaming of colliding the very-high-energy protons from the SPS with antiprotons to directly produce the bosons of the weak interaction, the $W^{\pm}$ and the Z^0, the latter being responsible for the weak neutral currents (Chap. 6). Using protons and antiprotons is a rather economic way to produce head-on collisions, because both types of particles can be stored in the same ring, just as electron-positron colliders store both types of particles in the same ring. The problem, however, were the antiprotons. They were created by blasting protons from the PS into a metal block and filtering out the antiprotons from the debris. This caused their beam size to be very large and unsuitable to produce high collision rates. Fortunately, Simon van der Meer had come up with a new scheme to reduce the beam sizes. He called it *stochastic cooling*.

Great Idea: Stochastic Cooling

By the time Schottky noise was observed in the ISR, electronic components had become sufficiently fast to construct a system that reduces the random motion of particles in a storage ring. This system must be fast enough to detect the transverse position of slices of the beam, illustrated as the blue blocks passing through the position sensor shown on the bottom of Fig. 9.4. Each slice performs oscillations until it reaches a fast, so-called *kicker magnet* that reduces the oscillations. On the way to the kicker, the signal from the sensor is amplified and manipulated in such a way that it removes the oscillations of the slices, one by one. This only worked, because the signal path through the

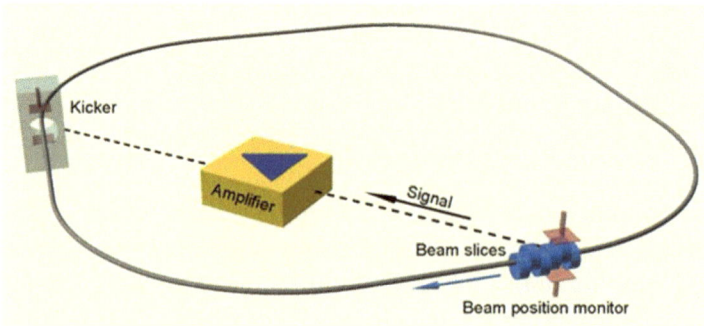

Fig. 9.4 Stochastic cooling

amplifier takes a shortcut along the sagitta of the trajectory that the beam takes, such that it is ready to excite the kicker appropriately to reduce the oscillation of just the right slice.

CERN Proton-Antiproton Collider

Figure 9.5 illustrates how the antiprotons needed to operate the SPS as a proton-antiproton collider, called Spp̄S, were prepared. After 26 GeV protons from the PS strike the antiproton-production target, antiprotons with an energy of 3.5 GeV were focused by a magnetic horn (Chap. 6) and collected in the *antiproton accumulator*, which has a circumference of 157 m. In this ring, several stochastic cooling systems reduced the beam sizes before the next pulse of antiprotons was received a few seconds later. Throughout the day antiprotons were produced and collected in the accumulator. Once a day's worth of antiprotons were ready, they were extracted from accumulator and returned to the PS, where their energy was increased to 26 GeV before they were transferred to the Spp̄S. Likewise, 26 GeV protons were directly transferred from the PS to the Spp̄S. Note that both in the PS and the Spp̄S protons move clockwise and antiprotons move counterclockwise. Three proton bunches and antiproton bunches each were injected into the Spp̄S which collided at six points in the Spp̄S.

When operating as a collider the Spp̄S increased the energies of both protons and antiprotons within a few seconds to 270 GeV at which point the magnets in the low-beta section (see Chap. 8) surrounding the collision points reduced, or "squeezed," the beam size to increase the collision rate. Once set up, the beams typically collided for up to 24 h at which point the beam intensity and beam quality had deteriorated, such that the Spp̄S had to be filled with fresh beams.

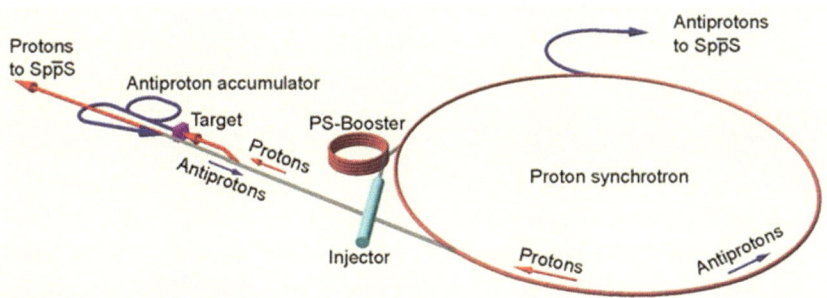

Fig. 9.5 Antiproton preparation for the proton-antiproton collider Spp̄S

One of the detectors, located in underground area 1 and hence called UA1, was huge: it was 10 m long, had a cross section of 6 m $\times$ 6 m, and a weight of 2000 tons. Much of the detector was permeated by a strong magnetic field in order to analyze the momentum of newly created particles. The trajectories of these particles could be reconstructed in three dimensions by analyzing discharges between almost 30 000 wires, many of them excited to high voltages. Surrounding this *central tracker*, but still embedded in the magnet, were components to detect the energy of particles, commonly referred to as *calorimeters*. Outside the iron yoke of the magnets, where only muons could arrive, more wire chambers tracked their trajectories. It is remarkable that the detector covered practically all directions such that no particles could escape undetected.

Since there were no other proton-antiproton accelerators around that could crosscheck anything that UA1 would find, the CERN management decided to fund a second detector, UA2, to crosscheck results. UA2 was less versatile than UA1, but also much less expensive. For example, the magnetic fields covered part of the detector. Instead it was optimized to very accurately determine the energy of the collision products, especially electrons and positrons.

$W^{\pm}$s and Z^0s

By 1981 both the Sp$\bar{\text{p}}$S was producing proton-antiproton collisions and the detectors were operational to start looking[14] for special decays of the charged carriers of the weak force, the W^+ and the W^-. The latter occasionally decays into an electron with a very high energy and a neutrino, which escapes undetected, but whose *missing energy* can be estimated from the other collision products. Towards the end of 1982 data were coming forth.[15] A fast system to rapidly identify promising-looking collisions, triggered a system to store all available data on magnetic tape. This trigger distilled the many billion collisions down to about a million promising ones. In this haystack of data, a meticulous detective work revealed five collisions in UA1 and four in UA2 that would pass all very strict acceptance criteria to qualify them as coming from a W^+ or W^-. This proved their existence. The next experimental run in the spring of 1983 provided a significant number of candidates for $W^{\pm}$, such that even their mass, around 80 GeV/c^2, could be determined.[16]

In the data from this run, both UA1 and UA2 found electron-positron pairs, whose energy added up to about 90 GeV, which was interpreted as evidence for the neutral carriers of the weak interaction, the Z^0. This proved the existence of the particle responsible for the weak neutral currents, discussed in Chap. 6. With the gauge bosons for the electroweak interaction discovered, evidence for the standard model was mounting.

A year later Rubbia and van der Meer were honored with a Nobel prize for the discovery of the $W^{\pm}$ and the Z^0 and for stochastic cooling which made this discovery possible[17].

Tevatron

Already during the construction of the Tevatron at Fermilab, plans emerged to use van der Meer's stochastic cooling systems to prepare antiproton beams and to convert the Tevatron to a proton-antiproton collider. This collider would be similar to the Sp$\bar{\text{p}}$S, albeit reaching much higher collision energies with beams of 900 GeV protons colliding with 900 GeV antiprotons. As in the Sp$\bar{\text{p}}$S, production of antiprotons was identified as the trickiest problem and the construction of the antiproton source started already in 1983; 120 GeV protons from the main ring were directed onto a copper target shown on the top left in Fig. 9.6. The antiprotons in the debris coming from the target were focused by a lithium lens. This is a short cylinder of lithium through which a huge electric current is passed to create a magnetic field that acts as a focusing lens. The antiprotons, having an energy of 8 GeV, were collected in a ring, called *debuncher* for technical reasons, and then transferred to a second ring, the *antiproton collector*. Stochastic cooling systems in both rings ensure adequate beam sizes for collider operation. Once sufficient numbers of antiprotons were available, they were transferred back to the main ring, where they circulated in the opposite direction of protons.

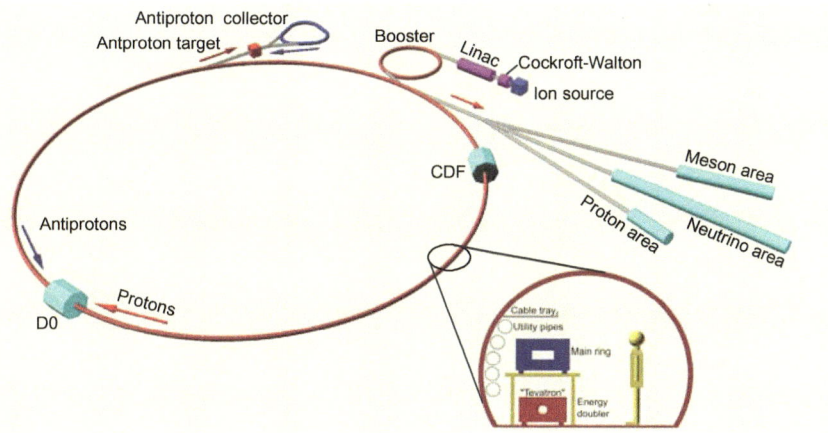

Fig. 9.6 Fermilab at the time of the Tevatron. Initially, only the main ring occupied the tunnel, the superconducting ring ("Tevatron"), the antiproton complex, and the detectors CDF and D0 came later

By 1987, the antiproton source was ready and connected to the main ring such that the Tevatron collider was ready to do physics. An operation cycle, commonly referred to as a *fill*, could start with creating protons in the ion source, accelerating them with the Cockroft-Walton, linac, and booster synchrotron to 8 GeV before injecting the beam into the main ring, which increased the energy to 120 GeV. At this point the beam was extracted and guided to the antiproton target, such that 8 GeV antiprotons could be accumulated. This process went on for several hours until a sufficient number of antiprotons were available in the accumulator. These antiprotons were partitioned into three bunches, transferred to the main ring, accelerated to 150 GeV, and then injected into the Tevatron. There, three counter-propagating proton bunches had been previously injected and were waiting to be joined by the antiprotons. Once both protons and antiprotons were stored, their energy was increased to 900 GeV which took about one minute. Then focusing magnets adjacent to the collision points squeezed the beam size at the collision point to increase the collision rate. In this configuration the detectors started to take data. This went on for hours until the beam quality and intensity had deteriorated. At this point the remaining beam was dumped and the magnets readjusted to injection conditions, such that newly prepared protons and antiprotons could be injected.

The *Collider Detector for Fermilab* (CDF) was the first detector that was ready to analyze the particles emerging from collisions. Like most modern detectors it was constructed like an onion with layers dedicated to specific tasks, such as determining the trajectories or the energies of escaping particles. Figure 9.7 illustrates the idea. To accomplish its task it had to be huge, about 10 m in each dimension and it weighed approximately 5000 tons. Closest to

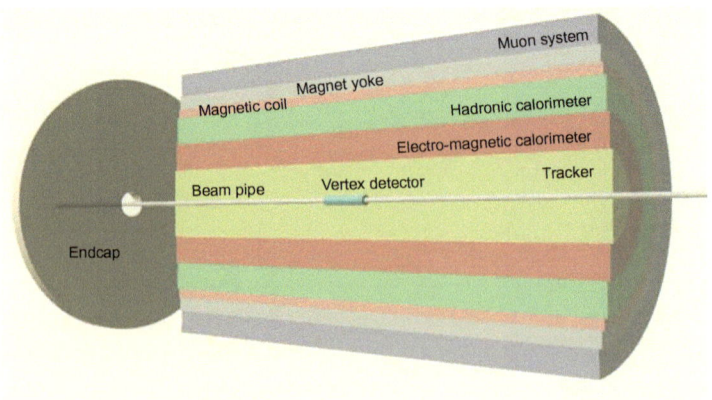

Fig. 9.7 Schematic sketch of a detector

the collision point the vertex detector consisted of semiconductor-based detectors, a precursor of the sensors in modern digital cameras. The next layer, the tracker, was based on drift chambers to determine the trajectories of particles. It was followed by the electro-magnetic calorimeter to determine the energy of escaping electrons and photons and many other low-energy particles that are stopped in it. The hadronic calorimeter then determined the energy of those particles that were not stopped in the previous layers. The magnetic coil and yoke were needed to create a strong magnetic field that bends the trajectory of escaping particles from which their momentum could be determined. Only muons made it past the calorimeter, the coil, and the magnet yoke. They were detected by the outermost layer, the muon system consisting of drift chambers. Finally, particles escaping in the direction of the beam pipe were detected in the endcaps on either side of the detector. CDF took several years to assemble and was ready to take data by 1985. A second detector, called D0 after its location in the ring, constructed similarly with central tracker, an elaborate calorimeter, and muon detectors, was operational by 1992.

Top Quark

The top quark had eluded all previous accelerators whose beam energy was just too low, but the Tevatron had a chance. The detectors were searching for specific decay modes of a top and an antitop quark. In one mode, the top quark decays into a bottom quark and a W-boson that immediately decays into two quarks that gives rise to two jets. The bottom quark has a relatively long lifetime and only decays into a jet after having traveled a little distance which is determined by the highly accurate semiconductor tracker. This identifies the jet as originating from a bottom quark. Likewise, the antitop decays into another antibottom quark that turns into a jet. It is identified by the small distance it had traveled from the point where the W-boson had decayed into a muon and a neutrino. The signature of a top-antitop meson is thus a muon and four jets, two of them displaced from the point of origin of the two other jets and the muon. By 1995, both CDF and D0 had sifted through the huge amount of data they had accumulated and identified about a dozen collisions each, having this and other but similar signatures. Moreover, the energy in the jets and the muon added up to indicate that the mass of the top quark is about 175 GeV. Since both detector collaborations independently and with a high level of confidence identified the top quark with a mass that was consistent with all theoretical calculations, this counts as its discovery.[18]

Tau Neutrino

Every few years the accelerators and detectors at Fermilab shut down for a longer period to install major upgrades. During times when only the detectors are pulled away from the beam pipe, the Tevatron could still produce protons for fixed-target experiments. During such a period, for a few months in 1997, the beams from the Tevatron bombarded a tungsten target. There it produced the neutrinos associated with the tau lepton that Martin Perl discovered at SPEAR (Chap. 8) more than 20 years earlier. After the target, magnets swept away all charged particles from the debris and a massive amount of shielding blocked all other particles. Only neutrinos would arrive at the detector located 36 m after the target. This detector for the *Direct Observation of Nu Tau* (DONUT) consisted of many layers of photographic emulsion interleaved with thin plates of stainless steel and with scintillating fibers that recorded the trajectories of charged particles coming from the interaction of a tau neutrino with the emulsion. Moreover, downstream of the emulsion calorimeters recorded the energy of emerging particles. Muon chambers even further downstream identified muons coming from the tau-neutrino interaction. Layers of electronic sensors recorded emerging electrons and muons and could pinpoint their point of origin in the emulsion to within about a millimeter. This made it possible to find that point after the emulsions were developed at a later time. The signature of a tau neutrino interacting in the emulsion is the track of a tau lepton that appears out of nowhere, because an invisible tau neutrino triggered it. After a very short distance the tau either decays into an electron or a muon that moves in a different direction than the tau, because undetected neutrinos recoil from the decay. Thus, the hunt for short millimeter-sized tracks with kinks was on. After digging through many thousand sites that were identified by the scintillating fibers and the other sensors, four candidate events remained that could only be explained by a tau neutrino causing it. Three years after the experiment, the DONUT collaboration[19] could finally announce its discovery.

Of course there were many more experiments done at the Tevatron, among them experiments to determine parameters related to CP violation and why our universe predominantly contains matter, rather than antimatter. Moreover, careful measurements of the masses and other properties of the W and Z bosons were done, even though most of the latter activities moved to the next generation of electron-positron colliders. They will keep us busy in the next chapter.

Notes

1. In 1974 the lab was renamed *Fermi National Accelerator Laboratory* and is usually referred to as Fermilab.

2. Wilson's cost cutting caused some headaches. The magnets of the main ring were designed a little too close to the edge and many developed electric shorts such that they had to be replaced. The problem was aggravated, because the magnets were installed during the winter and, once placed inside the warm and humid accelerator, they accumulated condensation water. All together this caused long delays during commissioning and later when operating the ring. Moreover, the beam pipes turned out to contain debris from manufacturing. Felicia, a trained ferret helped out and pulled a cleaning cloth through the beam pipe, but refused to crawl through long pipes. Cleaning was later taken over by a robotic device. More savings came from giving neither the experimental areas a concrete floor nor putting a solid roof over them. But this caused the experimenters major headaches (and cold feet) because their experiments were mounted on a mud floor with only a corrugated sheet-metal roof protecting delicate electronics from the harsh conditions during the mid-western winter. Wilson defended his frugal approach by arguing that any technology that worked the first time was over-designed and too expensive. Despite the savings, the budget was still large and Wilson had to defend it at a congressional hearing. When asked how accelerators help to defend the country, he famously argued that instead "they help make it worth defending." For a full account of Wilson's tenure and Fermilab's history see: L. Hoddeson, A. Kolb, and C. Westfall, *Fermilab, Physics, the Frontier, and Megascience,* University of Chicago Press, Chicago, 2008.

3. For an overview of the accelerators of this generation see: Robert Wilson, *The Next Generation of Particle Accelerators,* Scientific American, January 1980, page 42.

4. Early on during the experiment, the group found that the muon energies more often than expected added up to around 6 GeV. They first called this resonance "upsilon" and published their findings. Later, however, it turned out that the observed signature was based as a rare coincidence of random signals and disappeared in later studies. This "upsilon" thus was jokingly referred to as the "oops-leon." The signal that appeared at 9.5 GeV was nevertheless also given the name upsilon; this time the signal was solid and stood the test of time. For a vivid account by the discoverer himself see: Leon Lederman, *The Upsilon Particle,* Scientific American, October 1978, p. 72.

5. An account of the early developments is given in: J. Kunzler, M. Tanenbaum, *Superconducting Magnets,* Scientific American, June 1962, page 60.

6. Heike Kammerlingh Onnes (1853–1926) received the Nobel Prize in Physics in 1913 "for his investigations on the properties of matter at low temperatures which led, inter alia, to the production of liquid helium."

7. Absolute zero on the Celsius temperature scale is $-273\,°C$. In a scientific context, the Kelvin scale is used where absolute zero corresponds to $0\,K$.

8. Leon Lederman, *The Tevatron,* Scientific American, March 1991, p. 48.

9. Originally the ISR used conventional magnets only, but after 1978 superconducting magnets close to one interaction point were installed and increased the collision rates more than six-fold.

10. Already in 1973, two years after the first beam circulated in the ISR, the US initialized a design study for a beefed-up ISR, named *Isabelle.* Using superconducting magnets it was to collide $200\,GeV$ protons (and later even $400\,GeV$) to discover Ws and Z^0 bosons. Its construction started in 1978, but unfortunately, the chosen technology for the magnets was not sufficiently mature and caused delays. After the Ws and Z^0 were discovered at CERN, Isabelle became obsolete and the project was canceled. Two decades later, part of the Isabelle tunnel was filled with the *Relativistic Heavy Ion Collider* (RHIC). It also uses superconducting magnets and collides $250\,GeV$ protons or more often gold atoms.

11. The high collision energies put the ISR in a good position to make important discoveries, but the absence of a detector covering particles escaping in all directions prevented this. Instead, the ISR's legacy are the development of accelerator technology used in all later proton machines: Schottky noise diagnostic and stochastic cooling (this chapter), using the beam itself to ensure head-on collisions, superconducting magnets, vacuum technology, computer-based control of the accelerator, and storing enormous currents in the machine and reaching record collision rates. For a detailed account of the ISR, see: Arturo Russo, *The Intersecting Storage Rings: the construction and operation of CERN's second large machine and a survey of its experimental programme,* CERN-CHS-33, 1992. Available online from https://cds.cern.ch/record/232869.

12. The early days of touch screens are described in B. Stumpe, C. Sutton, *The First Capacitive Touch Screens at CERN*, CERN Courier, March 2010, https://cerncourier.com/a/the-first-capacitative-touch-screens-at-cern/.

13. Jointly with van der Meer, Carlo Rubbia (1934) received the Nobel Prize in Physics in 1984 "for their decisive contributions to the large project, which led to the discovery of the field particles W and Z, communicators of weak interaction."

14. David Cline, Carlo Rubbia, and Simon van der Meer, *The Search for Intermediate Vector Bosons,* Scientific American, March 1982, p. 48.

15. This highly competitive period in time is vividly described in: Gary Taubes, *Nobel Dreams,* Tempus books, Redmond, 1986.

16. Luigi Di Lella, *Remembering the W Discovery,* CERN Courier January 2023, https://cerncourier.com/a/remembering-the-w-discovery.

17. Up to this point in time, particle physics was dominated by labs in the US. The Nobel prize for Rubbia and van der Meer was the first one for the European labs, which now had three recent major discoveries under their belt: gluons at DESY, neutral currents and the Ws and the Z^0 at CERN. Always enjoying to report on a competition, the New York Times quipped "Europe three, U.S. not even Z-Zero". For the original article, see https://www.nytimes.com/1983/06/

06/opinion/europe-3-us-not-even-z-zero.html. For a more elaborate narrative, see: Gordon Fraser, *The Quark Machines,* Institute of Physics Publishing, Bristol, 1997.

18. Tony Liss and Paul Tipton, *The Discovery of the Top Quark,* Scientific American, September 1997, p. 54.

19. *DONUT comes to neutrino town,* CERN Courier, August 2000, available from https://cerncourier.com/a/donut-comes-to-neutrino-town/.

10

Particle Horns of Plenty

The large proton-antiproton colliders, Sp$\bar{\text{p}}$S and Tevatron, were very successful in discovering the heavy particles $W^{\pm}$, Z^0, and the top quark. Using heavy protons and antiprotons made it possible to reach unprecedented energies, but this success comes at a price. Protons are composed of quarks and gluons. In reality, not protons collide with antiprotons, but rather bags of quarks and gluons collide with other bags of quarks and gluons. Experimenters must spend an enormous effort to pinpoint just a few unique candidates that qualify as new particles. The debris rushing outwards from the collisions is just extremely difficult to interpret. But working out the details of sub-nuclear reactions requires a huge number of "clean" collisions between, preferably point-like, particles. This is just what electron-positron colliders, such as SPEAR, were doing. After all, as far as we know, electrons and positrons have no substructure and thus qualify as point-like. The logical next step was thus to build electron-positron colliders with much higher beam intensity and energy as their predecessors.

Already before the SPS was converted into a proton-antiproton collider, the CERN management planned the next accelerator, which, following the reasoning from the previous paragraph, was chosen to be a very Large Electron-Positron (LEP) collider.[1] The envisioned physics program focused on precision studies of the Z^0 and $W^{\pm}$ bosons that were earlier discovered in the Sp$\bar{\text{p}}$S (Chap. 9).

LEP

Already 1981, the CERN management had decided to build LEP[2] close to the CERN site in Geneva such that the PS and SPS could be used as pre-accelerators. But the accelerator had to be placed in an underground tunnel, 150 m below the Jura mountains in the west and Geneva airport in the east. After heroic negotiations with the funding bodies and landowners in Switzerland and France where parts of LEP was located, ground-breaking took place in 1983. The 27 km long tunnel with an inner diameter of a little under 4 m was bored in the following five years,[3] and in 1988 all components of the accelerator could be installed with the help of a monorail train, suspended from the ceiling of the tunnel. It transported all personnel and material from eight vertical access shafts to their destination in the tunnel. The large size of LEP required ingenious methods to ensure that the tunnel-boring machines actually meet at the end within a few centimeters. On this scale, the curvature of the earth becomes important. For example, the vertical direction, as defined by gravity, differs by a tenth of a degree from one side of LEP to the other.

Since the funding conditions were very tight, the scientists at CERN came up with ingenious tricks to save money. For example, due to the large size of the ring and the low mass of electrons the bending magnets that keep the beams on their circular path only had to provide rather low fields. By sandwiching concrete in-between the iron in the magnet yoke the cost of the magnets was halved. Considering that about 3400 magnets, each 5.7 m long, were needed; this constituted a considerable cost saving.

Replenishing the energy lost by the emission of synchrotron radiation required many radio-frequency accelerating cavities. During the first years of operation—usually referred to as LEP1—128 normalconducting copper cavities were installed. Being normalconducting they waste quite some energy in heating up the copper. In order to limit the losses in the cavities a second, much larger and less lossy, cavity was mounted on top of the accelerating cavities. It stored the microwaves for a few microseconds between the arrival of bunches in the cavity. This trick reduced the rather substantial electricity bill by about 40%.

Filling LEP with electrons and positrons was a highly orchestrated procedure. Electrons from a heated cathode were accelerated to 200 MeV and hit a tungsten target to produce positrons. The positrons were accelerated further to 600 MeV and collected in a small ring, the electron-positron accumulator. After sufficient numbers of positrons were available, they were injected into the venerable PS that had been upgraded with a special accelerating system for positrons and electrons. The PS then accelerated the positrons to 3.5 GeV before injecting them into the adapted SPS. The SPS brought the energy to 20 GeV at which point the beam was transferred to LEP. Apart from hitting the positron target, electrons follow essentially the same procedure: only each ring is circulated in the opposite direction. Finally four bunches of electrons and four bunches of positrons circulated in LEP. Once LEP was filled at 20 GeV, the energy was increased to about 50 GeV, the beam size at the four collision points was squeezed to increase the collision rate, and the four detectors installed in large underground caverns started taking data.[4]

The four detectors, each weighing several thousand tons, were constructed in layers, as discussed before. A silicon-based vertex detector, just outside the beam pipe and as close as possible to the collision point, was surrounded by a central tracker to reconstruct the trajectories of particles, followed by a layer that detects their energy. Usually this part was permeated by a magnetic field. A system to detect muons was placed further outside. Yet, each detector emphasized different aspects of the analysis. ALEPH (Apparatus for LEP Physics) was a general purpose detector that used a large gas volume superimposed by electric and very strong magnetic fields to reconstruct the trajectories, called a *time projection chamber* (TPC). DELPHI (Detector with Lepton, Photon and Hadron Identification) used large mirrors to collect so-called Cherenkov light from particles traversing a gas volume.[5] This made it possible to determine the speed of particles and to enhance the ability to characterize electrons and muons. L3 (Letter of intent number 3) used a large number of scintillating crystals to determine the energy and a large magnet to determine the momentum of the particles. OPAL (Omni-Purpose Apparatus for LEP) was a versatile detector, mostly based on previously established technology in order to ensure that one detector would be operational at startup though actually all four detector worked from day one.

Great Idea: World-Wide Web

The four collaborations operating the detectors had grown to include several hundred scientists and communication within such a collaboration required new tools. Around the time LEP started up, the CERN computing center, and in particular Tim Berners-Lee,[6] had come up with a new communication tool based on placing documents, marked-up with links to other likewise marked-up documents, on networked computers. Today we know these documents as *html* files and the servers as web servers. The protocol that governs the transactions is known as the *HyperText Transfer Protocol*, which explains the http prefix used to identify web pages.[7] Thus the infrastructure that soon became the *World Wide Web* was born at CERN out of the necessity to facilitate communication within large collaborations.

With the accelerator and the detectors operational and only a few weeks after injecting the first beams into LEP, the beams were accelerated to the energy corresponding to mass of the Z^0, and first collisions were observed. Already after a few minutes one of the detectors indeed reported the first Z^0. Shifts for running detectors and for improving accelerator performance were interleaved and after a few months, late in 1989, the collaborations published their first result in four back-to-back articles in a scientific journal.[8]

Three Generations

By measuring the production rate of Z^0s as a function of the beam energy the lifetime[9] of the Z^0 can be determined and that allows one to determine the number of possible ways Z^0s decay into lighter particles. Measurements from LEP and SLAC (see below) showed that the measured lifetime is accounted for by the known particles from the first three generations and this rules out that a fourth generation of leptons and quarks exists. We therefore only have to add one row to the table in Chap. 8 with two generations and arrive at table of particles from the inside of the front cover.

In the following years, while LEP operated with $50\,\mathrm{GeV}$ beams, the Z^0 resonance was thoroughly explored and many decay modes of the Z^0s were analyzed. The four detectors identified and reconstructed 17 million Z^0s which justifies LEPs characterization as a Z^0-factory. This large number of Z^0s made it possible to determine its mass as accurately as one part in $100\,000$. This precision relies heavily on the knowledge of the beam's energy. It was found to be affected by the tides due to the phase of the moon—even the New York Times[10] found this newsworthy—and the water level in Lake Geneva. Even passing high-speed trains affected the beam energy. The latter was noted, because the perturbations were absent during a railway strike. Moreover, comparing the production rate of electrons with that of muons and tau leptons, showed that all three were equal with high accuracy. This convincingly shows that these three leptons behave exactly the same, a fact dubbed *lepton universality*.

Since Wu's experiments in the late 1950s (Chap. 5) it was known that the violation of parity shows up as an asymmetry in count rates. Since the Z^0 is a force carrier of the weak interaction, some ejected particles are expected to show up more prominently in the downstream part of the detector, while others show up more often in the upstream direction. These so-called *forward-backward asymmetries* were carefully analyzed in all four detectors and validated the predictions of the standard model with high precision.

That gluons, the force carriers of the strong interaction, can interact among themselves is one of the key ingredients of the standard model. This self-interaction can be analyzed by comparing the number of events with two jets to those with three jets escaping from the interaction point. The predictions of QCD were confirmed and showed that the interaction strength gets smaller the closer the interacting particles approach each other. This is consistent with asymptotic freedom and the parton model (see Chap. 7) that is based on the assumptions that quasi-free quarks reside deep inside the proton.

In order to explore the $W^\pm$ bosons much higher beam energies were needed, because they had to be produced in pairs of a positive W^+ and a negative W^-. With their mass expected in the 80 GeV range, much higher accelerating voltages were needed and that required new technologies.

Great Idea: Superconducting Radio-Frequency System

At the foreseen beam energies the electrons lose energy by emitting copious amounts of synchrotron radiation that the accelerating cavities must replenish with high efficiency. In other words, the cavities must transfer all energy to the beam rather than losing energy by heating themselves up. Such losses are a bit like friction in the bearings of a swing with your niece on it. Lots of pushing is needed to keep your niece happy. On the other hand, greasing the bearings reduces the friction and allows your niece to enjoy much larger excursions. In the same way, reducing the losses in the cavities will allow the microwave generators to produce much higher accelerating fields.

Already in the late 1970s, CERN explored the feasibility of superconducting accelerating cavities to prevent these losses. This program was intensified during the first six years of LEP1 operation and by 1995 new cavities reaching several times higher fields than the old ones were produced on an industrial scale and available for installation. A year later, 160 superconducting cavities were in place, such that pair production of $W^\pm$ could start. This new phase of operation was referred to as LEP2 to signify the markedly increased performance of the accelerator. Until the year 2000 the number of superconducting cavities was almost doubled, which made beam energies exceeding 100 GeV possible. At this energy, *on every turn* in the accelerator, the particles radiate 3% of their total energy as synchrotron radiation. And this energy had to be replenished by the new cavities.

Once LEP2 was up and running[11] at the higher energies, the detectors started to explore the production of W bosons. They determined their mass to a high precision which was then used to perform consistency checks of the standard model. A thorough shake-down of many other features, predicted by the standard model, was done. For example, as expected from theory, it was convincingly shown that also the Z^0 can produce $W^\pm$-pairs. Many other measurements validated all predictions of the standard model with high precision. Since the model relies heavily on the renormalization procedure,[12] introduced by Gerard 't Hooft and Martinus Veltman,[13] even the underlying theory was validated. Confidence in the standard model grew.

At very high energies, the losses due to synchrotron radiation in LEP2 became prohibitive. Pointing two linear accelerators, one for electrons and one for positrons, at each other avoids this problem. Transforming the 3 km long linear accelerator at SLAC into the *Stanford Linear Collider* (SLC) provided a revolutionary way to explore this idea. It served both as a source for Z^0 bosons and as a prototype for future linear colliders.

Stanford Linear Collider

In order to operate with beam energies in the 50 GeV range the SLAC linac received an improved radio-frequency system with new power generators. These were equipped with ingenious circuitry that increased the peak power at the expense of the duration of the pulse. Now the linac was ready to accelerate both electrons and positrons to the energies needed to produce Z^0 bosons. But still there was only one linac available at SLAC, which therefore had to accelerate both types of particles before separating them in a beam switch yard and guiding them independently through their respective arcs to the detector. There they collided head-on and were subsequently guided to beam dumps.[14]

Figure 10.1 illustrates the basic layout and also the operation of the SLC. Two bunches of electrons, about 20 m apart, were produced in the electron source by shining a high-power laser onto a cathode, which knocks out several 10^{10} electrons. Both bunches are accelerated to 1.2 GeV before being injected into the electron damping ring where the emission of synchrotron radiation reduces the random motion of the particles and shrinks the beam sizes. Both bunches are then re-injected into the linac after a positron bunch, which was prepared in the positron damping ring. We get to where the positrons come

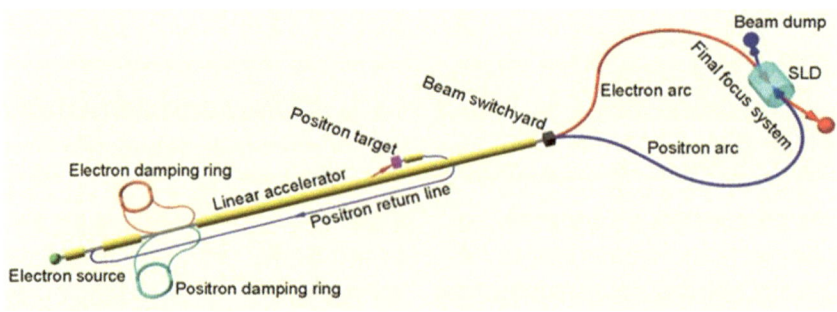

Fig. 10.1 Stanford linear collider

from shortly. All three bunches are then accelerated to about 30 GeV where the trailing electron bunch is directed onto a target to produce the positrons for the next pulse. The positron bunch and one electron bunch continue their journey through the remainder of the linac up to about 50 GeV. In the switchyard the electron and positron bunch are directed into their respective arcs, before an elaborate sequence of focusing elements—the final focus system—reduces their sizes to a fraction of a hair's width at the collision point. There they collide head-on in the middle of the SLD detector and end their lives in beam dumps. This cycle repeats 120 times per second.

Let us now return to the positrons created in the positron source. After the trailing electron bunch impinges on the target and creates the positrons, they are collected, accelerated to 0.2 GeV, and returned via the positron return line to the start of the linac. After injecting in the linac, their energy is increased to 1.2 GeV before they are injected into the positron damping ring where their beam size is reduced to make them ready for the next pulse.

Already in 1987 the accelerator was operational, but being the first machine of its breed, many "teething troubles" had to be understood and amended. This makes it clear that SLC's purpose was not only physics with Z^0s, but also to serve as prototype for a new class of accelerators, which required new methods to cope with perturbations of the micron-sized beams. Developing diagnostics of what's wrong with the beam and automatic correction algorithms played a key role to establish reliable operation. Moreover, new hardware and methods made it possible that eventually the beam sizes at the collision point became almost a hundred times smaller than the diameter of a hair. Since these tiny bunches carry opposite charges, they attract each other and one beam acts as a focusing lens for the other beam. With the help of this *pinch effect* the beams contract further. In some cases this doubled the collision rate.

Initially the upgraded Mark II detector, which had previously been used in SPEAR and in PEP, was installed at the collision point of the SLC and recorded the first Z^0s in 1989. In 1991 a new detector, the more modern *SLC Large Detector* (SLD), replaced it. As most detectors at the time, it sported a vertex detector close to the collision point, surrounded by tracker and calorimeters to determine the trajectories and energies of ejected particles. A Cherenkov detector, similar to the one used in the DELPHI detector at LEP, additionally measured the speed of escaping particles. A strong magnetic field that permeated the whole assembly made it possible to measure the momentum of the ejected particles. A large number of the approximately 600 000 Z^0s that SLD collected over the years originated from longitudinally polarized electrons.[15] This allowed the physicists to asses spin-dependent properties of the standard model.

This polarization was made possible by enhancing the capabilities of the electron source. In particular the laser that knocked out the electrons from a specially prepared cathode bequeathed its polarization to the electrons: either with spin parallel in the direction of motion, called right-handed, or anti-parallel, called left-handed. Using elaborate beam manipulations, this polarization could then be preserved all the way to the collision point. There the polarization was measured by bouncing a laser pulse off the electrons and determining the polarization of the recoiling photons. This polarization proved crucial to work out whether left-handed electrons produce more ejected particles of some kind than right-handed electrons do and thereby reveal subtle features of the weak interaction that are complementary to those found at LEP.

All detectors in LEP and SLC featured silicon-based vertex detectors that surround the beam pipe near the collision point and could determine the point of origin of trajectories with micron precision. This made it possible to identify, for example, bottom quarks which live long enough to travel a short distance before giving rise to a jet. If this jet appears from a second point of origin, different from the point of the primary collision, it was likely from a decaying bottom quark. Selectively analyzing the decay modes of bottom quarks thus became possible. As a matter of fact bottom quarks are prime candidates to understand CP violation, the imbalance between matter and antimatter in our universe.

CP Violation Revisited

As discussed in Chap. 6 CP violation was first observed in the weak-interaction decay of neutral K-mesons by Cronin and Fitch in the early 1960s. About ten years later, Makoto Kobayashi and Toshihide Maskawa[16] worked out a way of how to accommodate this effect into the emerging standard model for quarks. They found that the math only worked if there were three generations with a total of six flavors of quarks. At the time, however, only three flavors were known—up, down, and strange—so their theory remained unnoticed. Their luck turned when the fourth, the charm quark, was discovered (Chap. 8) in the following year. Moreover, two years later, the tau-lepton as the first member of the third generation of elementary particles was found, followed by the next member of the third generation, the bottom quark (Chap. 9).

In the original framework for the weak interaction, developed by Glashow, Salam, and Weinberg, a down quark can only decay into an up quark but never into a charm or a top quark. In Kobayashi and Maskawa's theory, on the other hand, the three quarks with charge $-1/3$ (down, strange, bottom) appear as quantum mechanical mixtures and can decay,[17] with certain probabilities, into one of the quarks with charge 2/3. Either an up, a charm, or a top quark can appear. This behavior is referred to as *flavor* or *quark mixing*. The mixing of the K-mesons, discussed in Chap. 6, is an early example though there the effect is very small. Later, however, experiments with upsilon mesons (Chap. 9) in the DORIS collider at DESY showed that mixing of bottom quarks is much more pronounced.

Unfortunately, the processes with bottom quarks are extremely rare, such that enormous collision rates are needed in order to explore the different ways that bottom quarks decay and the probabilities that they occur. One way of making these rare events stand out is by displacing them from the collision point. And that can be done by giving the electrons and positrons different energies and storing them in separate rings.

Great Idea: Double-Ring Colliders

But this requires two rings operating at different energies—one for electrons and one for positrons—that intersect in one or several collision points. Normally the electrons are stored in the higher-energy ring. With such a setup collision products that have a longer lifetime, such as the mesons with the bottom quark, move forward in the same direction as the beam with the higher energy. Thus, their point of decay is clearly discernible in a high-resolution vertex detector that surrounds the collision region and makes identification of interesting events possible. But we also have to increase the collision rate.

That is made possible by filling each ring with hundreds or even thousands of bunches. Storing electrons and positrons in separate rings avoids unwanted collisions outside the detector that perturb the beams. It was, however, difficult to maintain a constant and high beam intensity, because particles are lost in the collisions with the other beam and with unwanted collisions with gas molecules. Monitoring the intensity of the bunches and selectively topping them off in what was called *trickle injection*, the collision rates were constantly kept at a high level.

But with the increased beam intensities new problems appeared. The many high-intensity bunches in the same ring perturbed each other by exciting fields in the beam pipe that later bunches "feel". The process can cause the many bunches to become unstable, much like a microphone picking up noise from a loudspeaker leading to an annoying whistle. Developments of fast electronics in the 1990, such as digital signal processors, came just in time to be used for feedback systems to stabilize the beams.

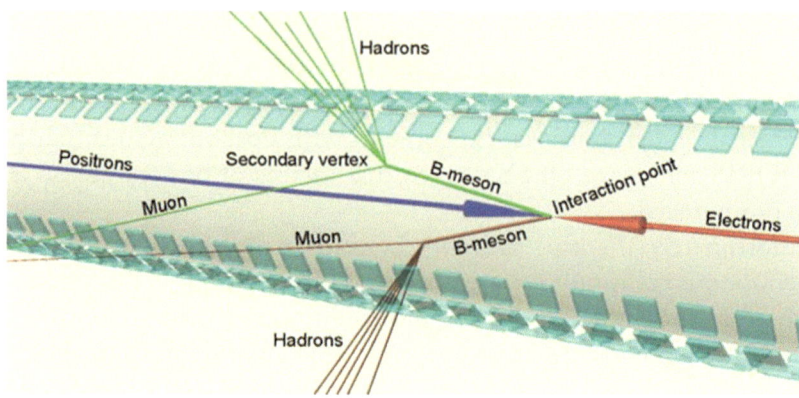

Fig. 10.2   Vertex detector

Two double-ring colliders, so-called *B-factories*, were constructed in the 1990s, one at the KEK laboratory in Japan and a second one at SLAC in the US.

B-Factories: PEP-II and KEKB

At SLAC, the 15 GeV collider PEP with a circumference of 2.2 km was refurbished and turned into the PEP-II B-factory. The beam energies (9 GeV electrons and 3.1 GeV positrons) were chosen to create an excited state of the upsilon meson that rapidly decays into two neutral B-mesons which consist of a bottom quark and a down quark. The point of decay was pinpointed by the silicon-based vertex detector in the *BaBar* detector, as shown in Fig. 10.2, and the ejected particles were analyzed further in its outer layers. It is remarkable that the accelerator performed so well that the amount of data exceeded the capability of the on-site computing center. As a remedy, the test infrastructure for the LHC data acquisition (Chap. 11) was copied and data were sent to participating laboratories via high-speed networks—one of the first high-volume applications of *grid computing.*

In Japan, the 32 GeV electron-positron collider Tristan with a circumference of 3 km was converted to the *KEKB* B-factory with its detector named *Belle.*[18] The features of this collider are rather similar to those of PEP-II. By 1999 the detectors in PEP-II and KEKB started to take data and by 2008 had accumulated about a billion events featuring B-mesons, making high-precision measurements of the decay probabilities possible and validating Kobayashi's and Maskawa's theory of quark mixing. All relevant mixing rates among the quarks were found with high precision. This allowed the physicists to explore small discrepancies in the standard model, though at the time of writing, none had been found. The standard model passed all tests with flying colors.

The B-mesons proved fruitful, but even the kaons that puzzled Cronin and Fitch in the first place (Chap. 6) benefited from the boost in accelerator technology. In the year 2000 the electron-positron collider *Daphne* started operation.

Phi Factory: Daphne in Frascati

Like the B-factories Daphne consists of two interleaved rings, each having a circumference of 98 m. Only here the beam energy of 510 MeV is the same in both rings. A large number of bunches provides high beam intensities and collision rates, which makes it possible to measure small differences in decay rates with high precision. The machine is fine-tuned to produce copious numbers of phi mesons that contain a strange quark and its antiquark. These phi mesons decay about a third of the time into two neutral kaons, a short-lived K_S and a long-lived K_L. While the K_S quickly decays inside the beam pipe, the K_L and their decay products are observed in the surrounding detector, called *KLOE*. From the different decay modes subtle features of CP violation are extracted. Since 2014 an upgraded detector *KLOE-2* is operating and collecting more than a billion kaon events, promising even higher precision in the future.

With the discoveries from the large proton machines and the precision results from the electron colliders the standard model was in great shape. It provided a consistent framework that described all known fundamental particles—quarks and leptons—and the forces—electro-magnetic, weak, strong. There was, however, one piece missing, the *Higgs boson*. It had been introduced in the early 1960s as part of a theory to explain the masses of elementary particles. Later it proved to be essential for the renormalization procedure (see Chap. 8) that renders the theory mathematically consistent.[19] Towards the end of the first decade of the 21st century, however, the Higgs particle had still not been observed. It was time to change that. Already in the mid-1980s three new accelerator projects set out to find it.

Notes

1. A first-hand account of LEP's history is given by its director general at the time: Herwig Schopper, *LEP—The Lord of the Collider Rings at CERN, 1980–2000*, Springer Verlag, Heidelberg, 2009.
2. The project leaders themselves describe LEP in: Stephen Myers and Emilio Picasso, *The LEP Collider*, Scientific American, July 1990, page 54.

3. Before the channel tunnel connecting France and the UK, the LEP tunnel was the longest tunnel in Europe. In order to avoid excessively deep access shafts under the Jura mountains and the ground water level near Geneva airport, the LEP tunnel is inclined by 1.8 degrees. Digging it with three tunnel-boring machines was far from trivial. Especially, establishing a unique coordinate system above and below ground required state-of-the-art survey methods and equipment. Another difficulty came from a major ingress of water under the Jura mountains, which delayed the project by six months.

4. The energy available to create new particles is then 100 GeV, the sum of the beam energies of electrons and protons.

5. Particles moving faster than light in a medium emit radiation which is called Cherenkov light.

6. While working at CERN, Tim Berners-Lee (1955) invented the infrastructure and the network protocols that made the world-wide web possible that we today call *Internet*. Today he is the director of the *World-Wide Web Consortium* (W3C) which oversees the development of the Internet.

7. The first proposal describing the protocol used on the world-wide web is available from https://www.w3.org/History/1989/proposal.html. For a description of the early heydays, see: Brian Hayes, *Computing Science: The World-Wide Web,* American Scientist, September 1994, page 416.

8. The four LEP detector collaborations published their results in the journal Physics Letters B231 (1989) starting on page 509.

9. Actually the width of the resonance is measured. It is inversely proportional to the lifetime. For more background, see Gary Feltman and Jack Steinberger, *The Number of Families of Matter,* Scientific American, February 1991 page 70.

10. New York Times article on Nov 27, 1992 "Moon is blamed for blips in a particle accelerator."

11. After installation of the superconducting cavities was complete and beam was injected into LEP, the beam refused to go around the accelerator. The operators heroically tried to direct the beam around some apparent obstacle in the beam pipe, which, after several days of effort, turned out to be two Heineken beer bottles that somebody had stuffed into the beam pipe. After they were removed and some additional cleaning, the beam actually made it around LEP. For a first-hand account of this incident, see: Steve Myers, *The Greatest Lepton Collider,* CERN Courier, September 2019. Available online from https://cerncourier.com/a/the-greatest-lepton-collider.

12. Gerard 't Hooft, *Gauge Theories of the Forces between Elementary Particles,* Scientific American, June 1980, page 104.

13. Gerard 't Hooft (b. 1946) and Martinus Veltman (1931–2021) received the Nobel Prize in Physics in 1999 "for elucidating the quantum structure of electroweak interactions in physics."

14. For a first-hand description of the SLC see: John Rees, *The Stanford Linear Collider,* Scientific American, October 1989, page 58.

15. In longitudinally polarized beams the spin of the electrons is either parallel or anti-parallel to the direction the electron travels. This type of polarization is particularly useful to determine spin-dependent effects. Also in LEP the beams were polarized, but in the direction perpendicular to the direction of propagation. These transversely polarized beams played a lesser role for the experiments, except they made it possible to determine the beam energy with exceptional precision. This was actually used to observe the influence of the moon, mentioned in a previous section of this chapter.
16. Makoto Kobayashi, (b. 1944) and Toshihide Maskawa (1940–2021) received the Nobel Prize in Physics in 2008 "for the discovery of the origin of the broken symmetry which predicts the existence of at least three families of quarks in nature."
17. Apart from the quark, a lepton and its neutrino appear in this decay.
18. The bottom quark was sometimes also called "beauty," which is "belle" in the French language.
19. Martinus Veltman, *The Higgs Boson,* Scientific American, November 1986, page 76.

11

Large Hadron Colliders

Already before the 27 km tunnel for LEP was dug, folks at CERN dreamt of using the tunnel for a very-high-energy proton accelerator. Colleagues in the Soviet Union and the US had similar dreams and they even had a head start.

In the early 1980s the Soviet Union started constructing an accelerator, called the *UNK*, in Protvino near Moscow. A 21 km long tunnel was to contain a fairly conventional synchrotron with normalconducting magnets. It would first accelerate protons to 400 GeV and then inject them into two superconducting rings, which would accelerate them further to 3 TeV where they collide with protons stored in a second ring. Following the collapse of the Soviet Union, funding for the UNK stopped and the project was abandoned.

In the US, design work for the *Superconducting Super-Collider*[1] (SSC) got under way in 1984. After five years a site near Dallas in Texas was selected where construction of the project started in 1989. Three progressively larger synchrotrons would accelerate protons to 2 TeV and then inject them into two counter-propagating rings with superconducting magnets. There two beams of protons would be accelerated to 20 TeV in a 87 km long underground tunnel and then collide inside the two detectors SDC and GEM. Unfortunately, massive cost increases led to the project's termination in 1993, leaving a partially dug tunnel behind.

At CERN, in parallel to building LEP in the 1980s, the technical design of a future proton collider in the LEP tunnel got under way. After almost a decade of progress, the project had reached a mature state. At this point the CERN member states were approached to decide about the funding. Despite difficulties to secure this funding, especially Germany struggled with the large costs of reunification after the fall of the Berlin wall, by 1994 the CERN

© The Author(s), under exclusive license to Springer Nature Switzerland AG 2024
V. Ziemann, *Beams*, Copernicus Books,
https://doi.org/10.1007/978-3-031-51852-2_11

member states reached a consensus and authorized the construction of the *Large Hadron Collider* (LHC).[2]

LHC

The LHC[3] was installed in the LEP tunnel after the scientific program of LEP ended in 2000. Using superconducting magnets that produce 60% higher fields than any other previously manufactured accelerator magnet allowed LHC to accelerate protons to an energy of 7 TeV within the available tunnel. This is only about a third of the energy that was foreseen at the SSC. To make up for the lower energy, the LHC aimed to achieve collision rates at least ten times higher than the SSC. To do so, LHC would collide two counter-propagating beams of protons rather than protons and antiprotons, because producing the large number of antiprotons to fill the 27 km ring was deemed impossible. But two proton beams require magnetic fields of opposite polarity and therefore cannot be stored in the same magnet structure as had been done in the Sp$\bar{\text{p}}$S or the Tevatron. Thus, two rings were needed, one for protons going clockwise and one for protons going counterclockwise. The problem, however, was that the tunnel was not wide enough to accommodate two separate rings with rather bulky superconducting magnets.

Luckily already in the 1970s, a solution to this dilemma was invented. The idea was based on placing two magnets with opposite polarity into one mechanical structure, which was called *two-in-one magnet.*

Great Idea: Two-in-One Magnets

The bending magnets, their cross-section is shown on the sketch in Fig. 11.1, are the heart of the LHC. The magnetic fields to keep the protons inside the beam pipe on their 27 km long circular journey are excited by the coils made of superconducting wire, shown as red crescents. The distance between the two beam pipes holding the counter-rotating beams is a little under 20 cm. In order to guide the beams for many hours in LHC, extreme demands are placed on the field quality and thus on the placement of the coils. This is achieved by tightly pressing them into shape with an aluminum collar, shown as the darker grey rectangle surrounding the coils. The collar is then embedded in an iron yoke which shields the outside from stray magnetic fields. The yoke and everything inside is then cooled by liquid Helium and is referred to as the *cold mass*. The surrounding shrinking cylinder presses the whole assembly into its final form and thereby fixes the coils in their final position. The cold mass is

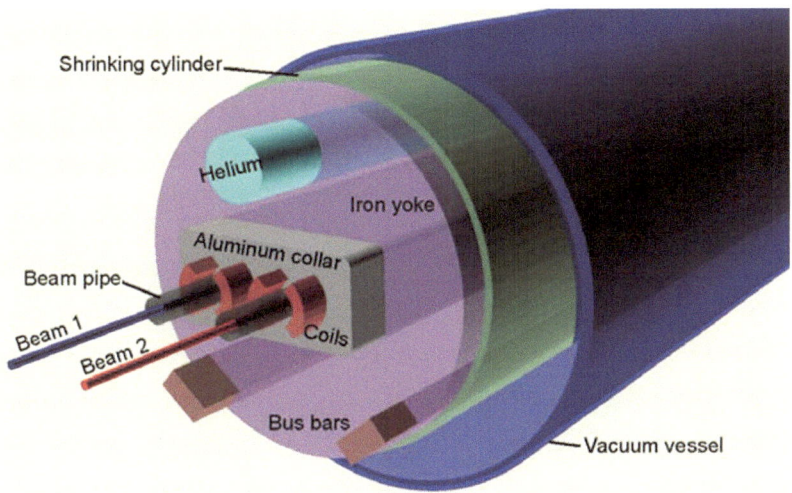

Fig. 11.1   Two-in-one magnet for LHC

surrounded by a vacuum vessel, which works much like a thermos bottle; it keeps the inside cold by preventing outside heat from entering the cold mass. Each of these bending magnets is 16 m long and weighs approximately 30 tons. They fill about three quarters of the length of the LHC tunnel.

Superconducting magnets make the high fields possible but only up to a limit at which the fields themselves destroy the superconductivity. A way out of this dilemma is cooling the magnets to lower temperatures which extends the limit to higher fields; in LHC the temperature is reduced from the previously-used 4.5 K down to 1.9 K. At this temperature liquid Helium becomes superfluid which exhibits additional benefits. It loses all its viscosity and creeps into the magnet windings where it efficiently absorbs heat. A high heat-carrying capacity and frictionless motion through the long pipes then quickly removes the heat from the magnet and the superconducting coils. Frictionless motion is important, because the magnets in one octant of LHC, shown in Fig. 11.2, are connected in series into one 3.5 km long continuous thermos bottle, through which the liquid Helium must move and cool down several thousand tons of cold mass.

The magnets in one octant guide the protons on arcs that connect eight straight sections where the experiments and other components to ensure safe operation of LHC are located. The straight sections are numbered in the clockwise sense from one near the ATLAS detector to eight near LHCb. The two proton beams, shown in red and blue in Fig. 11.2, arrive from the SPS and are injected into straight sections two and eight, just upstream of the detectors

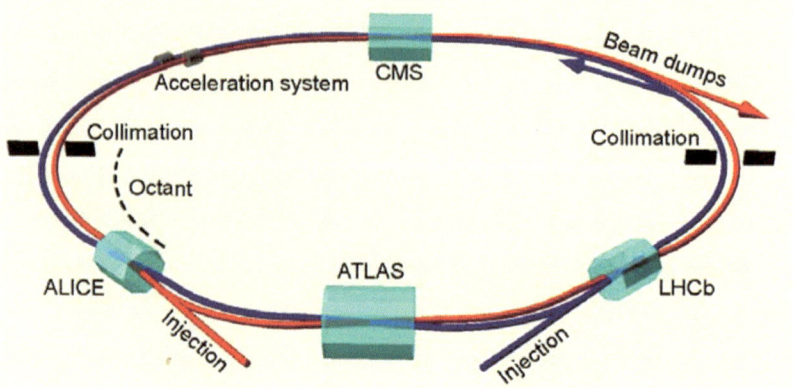

Fig. 11.2   Large hadron collider

ALICE and LHCb. The two beams collide inside the four large experiments ATLAS, CMS, ALICE, and LHCb where they cross from the inside to the outside. Straight section four houses radio-frequency cavities to accelerate the beams. These cavities are also superconducting in order to minimize losses and thereby reduce the electricity bill.

The energy stored in the beams corresponds to thousand tons of water falling form a church tower. This can seriously damage the accelerator if lost in an uncontrolled way all at once, but even small beam losses heat up the superconducting magnets such that they become normalconducting and stop deflecting the beams. Three straight sections are therefore devoted to prevent this calamity. Number three and seven contain so-called *collimators*—movable blocks of material that intercept protons that have strayed too far from the center of the beam pipe. In this way they are lost in a controlled way in designated places, without heating magnets. If collimation is insufficient or some essential component in LHC malfunctions, the beams are kicked out of the ring and directed to beam dumps where they are safely disposed of.

A typical day in the life of LHC begins by burning hydrogen gas in a plasma, extracting the positively charged protons, accelerating them in a sequence of smaller accelerators before injecting them into the proton synchrotron (Chap. 6). Once the protons are accelerated to 26 GeV, they are transferred to the SPS where their energy is increased to 450 GeV. At this point the protons are transferred to the LHC. Since the circumference of the SPS is much smaller than that of the LHC, only a fraction of the LHC circumference is filled. Therefore this procedure is repeated until most of the circumference is filled and close to 3000 bunches circulate in each of the two rings. From this point it takes about 20 minutes to accelerate the protons to the peak energy of close to 7000 GeV or 7 TeV. Now the beams are squeezed to small sizes inside the detector and data taking begins. The experimenters then continue to take data until the beam intensity has deteriorated significantly. At this point the beams are directed to the beam dumps and the magnet excitation is lowered to the value needed to start a new injection. This operation cycle typically repeats twice a day.

ATLAS and CMS

After the demise of the UNK and the SSC it was obvious that LHC would be the only accelerator providing multi-TeV beams. In order to provide independent confirmation of discoveries two large detectors, exploiting somewhat different methods to identify particles, are installed in LHC. Straight section one houses *A Toroidal LHC ApparatuS* (ATLAS) and straight section five houses the *Compact Muon Solenoid* (CMS). Both detectors are much larger than their predecessors. ATLAS, a bit larger than CMS, is almost 50 m long and has a diameter exceeding 20 m. CMS, on the other hand, is a bit heavier and weighs in at 12500 tons. ATLAS uses a magnetic field that winds around the direction of the proton beams—it's called toroidal—whereas CMS uses a field that is parallel to the direction of the protons—it is called solenoidal.

Despite using different technologies for subsystems both detectors follow the onion-paradigm from Fig. 9.7 with silicon-based vertex detectors closest to the collision point and just outside the beam pipe. The trackers in the next layer determine the trajectories of charged particles. Being rather close to the collision point, these detectors are custom-designed to withstand enormous levels of radiation. The next layers house calorimeters to measure the energy of electrons and photons, followed by a layer to detect the energy of hadrons. The magnetic fields permeate the entire central region of the detector and curve trajectories, which allows the physicists to determine the momentum of particles. Furthest outside trajectory trackers for muons are installed. Both ATLAS and CMS are hermetically sealed; they intercept particles escaping in all directions. This allows the identification of undetected neutrinos from a mismatched momentum and energy balance of reactions.

As a consequence of the rapid development of electronic components, practically all subsystems in the LHC detectors are more performant than those in LEP. In particular, the spatial resolution of the detectors is much higher as the pixel size of the vertex detectors has decreased significantly. This led to a dramatic increase of the amount of data that each collision provides. More than 100 million channels, of which about 80% come from the high-resolution vertex detectors, need to be recorded for *one* collision.[4]

Considering that the close to 3000 bunches in each beam collide 40 million times per second, the amount of data is staggering. You can visualize each detector as a 100-megapixel camera recording 40 million pictures each second. This amount of data is impossible to store and therefore ultrafast selection algorithms, so-called *triggers*, only retain data from events that look vaguely interesting. Since this decision takes some time, the data are fed into a storage pipeline, from which only the interesting events are pulled and passed via optical fiber links to a processor farm which reduces the rate further. In the end, less than one thousand events (with 100 million channels each) per second are retained. Instead of continuously filling terabyte-capacity harddisks, data are shipped via high-capacity computer networks to collaborators around the world where the data are analyzed—that's *grid computing*. From these developments to cope with the vast amount of experimental data at CERN, commercial vendors like Amazon or Microsoft developed products that go by the name of *cloud computing*.

The ATLAS and CMS collaborations constituted themselves around 1995 and had grown to 3000 members each by the time the detectors were assembled and ready to take data in 2009. Their main task was, of course, to find the Higgs boson.[5]

Finding the Higgs

So how does one find a Higgs boson? One has to diligently sift through petabytes—that's millions of giga-bytes—of data to detect the signature of a particular decay mode that stands out from many other decay modes. The signature ATLAS and CMS were mostly looking for is the decay of a Higgs boson into two high-energy photons, whose energies reveal the mass of the originating Higgs. Unfortunately, events with this particular signature are few and far apart. In the data set covering two six-month periods of running in 2011 and 2012, only about 200 000 Higgs bosons are hidden, of which 600 decay into two photons. These Higgs-related events show up as a small bump on top of a large number of other events that produce two photons. As it turned out, all these Higgs events assemble around one particular energy. Moreover, both ATLAS and CMS *independently* found this enhancement *at the same energy* of 125 GeV. An elaborate statistical analysis of the signals showed that the small bump stuck out above all natural fluctuations more than five times above the noise level. Physicists call this a five-sigma signal where sigma denotes the noise level. Five sigmas gives the bump a one-in-three-million chance that it comes from an "accidental conspiracy" of fluctuations. And a one-in-three-million chance to be wrong is commonly accepted as sound proof that the signal is real. Consequently, the ATLAS and the CMS collaborations claimed to have discovered "a particle consistent with the Higgs boson" at the press conference mentioned in the introduction.[6]

These claims of the detector collaborations were corroborated by the analysis of several other decay modes, such as a Higgs producing two Z^0 bosons, each of which decays into a muon pair or an electron pair that shows up in the detectors. In order to establish trust in the detector, both ATLAS and CMS had earlier performed a thorough shakedown of the standard model by producing and identifying all particles generated in earlier accelerators. This included the J/ψ (Chap. 8), bottom and top quarks (Chap. 9), as well as Z^0 and $W^{\pm}$ bosons (also Chap. 9). All experiments confirmed calculations based on the standard model within the measurement tolerances, giving great confidence, both in our understanding the model, as well as understanding the behavior of the large detectors with their many different subsystems. With the well-understood detectors other decay modes of the Higgs revealed even more of its properties, such as the spin. In this way it became certain that the particle with a mass of $125\,\mathrm{GeV}/c^2$ is indeed the Higgs. Finally, the last missing particle of the standard model was found which, in some sense, completes its description. So, the LHC delivered what was ordered, so what is left to do?

Well, answering open questions and solving puzzles is left to do! First, many free parameters of the theory, such as the masses of fundamental particles, must be specified, so-to-speak, by hand. They do not emerge self-consistently from within the theory. Second, the standard model only covers three of the fundamental forces (electromagnetism, weak and the strong interaction), but not gravity. Third, the formation of hadrons from quarks and gluons (hadronization) is not properly understood. Fourth, the preponderance of matter in our universe is still not fully explained. Fifth, modern developments in astronomy have shown that the particles of the standard model only correspond to about 5% of the energy in the universe. About 25% make the stars in galaxies move faster than observable matter can explain; it is called *dark matter* and we have no clue as to what it is. Even worse, 70% make the universe expand at an increasing rate. It is called *dark energy* and we have even less of a clue about this one. Clearly there must be more in this world than the standard model can describe.

Beyond the Standard Model

With so many open questions and puzzles one of the great hopes of LHC and the detector collaborations is to find cracks in the standard model which indicate new phenomena. They are commonly called "physics beyond the standard model." One way to identify these cracks is to measure phenomena predicted by the theory and hope that theory and experiment differ by more than the expected tolerances.[7] Another way is to search for particles that are not part of the standard model. A bit distressingly, so far the standard model rules and neither discrepancies between experiments and theory nor new so-called *exotic particles*, for example predicted by a theory called *supersymmetry*,[8] have turned up. It is a bit like a feeling: we know there must be something more, but we cannot yet find a handle to figure out what it is.

But we keep trying. Beside ATLAS and CMS, two of the other larger experiments in LHC are probing the details of the standard model in the hope to find cracks or to understand new features of the theory. LHCb takes a closer look at the preponderance of matter in our universe.

LHCb

Like the detectors in the B-factories (Chap. 10) LHCb is devoted to analyze particles containing bottom quarks which motivate the letter "b" in its name. At the high LHC energies these quarks are ejected close to the direction of the proton beams. The LHCb detector therefore extends about 20 m in the direction of one of the proton beams to catch and analyze the ejected particles close the beam pipe. The b-mesons created in a collision live long enough to move by a short distance before they decay and this identifies them as b-mesons. Once the proton beams circulate stably, a vertex detector is moved within a few millimeter of the collision point, allowing the detector to pinpoint this secondary vertex with micron precision.

Besides contributing to a better understanding of CP violation, LHCb found a large number of new particles made of previously unseen combinations of quarks, such as the Ξ_c. This is a combination of a charmed, a strange, and a down quark. So far it was always assumed that quarks either combine in groups of three to form baryons, or in pairs to form mesons, but LHCb found a number of particles composed of four and even five quarks. These states are referred to as tetraquarks and pentaquarks, respectively.

The ALICE experiment, on the other hand, explores the creation of matter immediately following the Big Bang by creating mini-big-bangs when colliding lead ions made of 208 nucleons each.

Lead Beams

After being created in an ion source, lead beams are accelerated in a dedicated linear accelerator, a small ring needed to reduce their beam size, the PS and the SPS, before they are injected into the LHC. In the LHC a little less than 600 bunches with moderate intensity circulate in each ring, such that their collision rate is much lower than that of proton beams. But being composed of 126 neutrons and 82 protons all hell breaks loose when two lead ions do collide head on. They create a *quark-gluon plasma* where the protons and neutrons lose their identity and create a high-density soup of quasi-free quarks and gluons, a state that resembles our universe about a millionth of a second after the Big Bang.[9] As the plasma expands, it cools down and the quarks and gluons combine to form three-quark baryons and two-quark mesons that rush outwards. Up to 20 000 particles can emerge from such a collision, but the moderate beam intensities limit the number of these events. Clearly, these enormous numbers of particles—more than a hundred times more than appear in proton-proton collisions—require a special detector named *A Large Ion Collider Experiment* (ALICE) to cope with them all.

ALICE

ALICE uses a four-layer vertex detector to pinpoint the starting point of the many trajectories and a volume, several meter in diameter, where the escaping particles knock out electrons from a gas. Measuring the arrival time and the position of these electrons in a detector called *Time-projection chamber* (TPC) the trajectory of the particles is reconstructed. Remarkably, this works even for thousands of particles. Other detectors measure the arrival time of particles with high precision and thereby deduce their speed whilst their energy is determined in so-called calorimeters, populating the outer layers of the detector. By correlating the information from the subsystems the type of particle—whether it is a kaon, a pion, or a proton—is determined.

By counting the number of particular species emerging from the collisions the physicists learn about the emergence of matter immediately following the Big Bang. One particular feature is that the originally free quarks and gluons get stuck in the quark-gluon plasma and have a hard time to form high-energy jets of hadrons, a phenomenon called *jet quenching*.

Up to this point we discussed the LHC as it was operating up to the time of writing (2022). But there are upgrades planned that will extend its productive lifetime for another 20 years, at least.

High-Luminosity LHC

Increasing the number of collisions is always welcome and therefore the *High Luminosity LHC*[10] (HL-LHC) upgrade aims to increase the collision rate at least fivefold. The two main points are doubling the beam intensity in LHC and making the beams much smaller at the collision point. Increasing the intensity requires a major upgrade of all pre-accelerators from the linear accelerators, to the PS booster, the PS, and the SPS. More beam in LHC also means more stray particles that can be lost in the superconducting magnets which requires an upgrade of the beam cleaning system. Making the beam sizes at the collision points smaller requires new focusing magnets with significantly increased strength. Moreover, to avoid that the closely spaced bunches collide outside the detector, the beams will no longer collide almost head-on, but at a larger angle, which, unfortunately, reduces the collision rate. Luckily, installing special hardware, so-called *crab-cavities,* will alleviate this problem. All detectors in LHC will also be upgraded to be able to cope with the much-increased collision rate and consequently many more particles that come from the collisions.

Preparations for all upgrades are ongoing and will be installed in the next longer maintenance shutdown around 2026. HL-LHC will then start taking data by 2029 and continue to do so for another decade, at least. So, what comes after 2040?

High-Energy LHC

The fields in LHC magnets are the highest that could be reached with the superconducting wires that were available at the time LHC was constructed. In the meantime, however, new materials appeared that hold the promise to double the achievable fields and thereby the peak energy in LHC. The *High Energy LHC* (HE-LHC) project now aims to develop the magnet technology that makes industrial and economic production of such magnets possible. If this technology matures sufficiently, it provides a way to extend the lifetime of LHC even beyond 2040.

But these upgrades of LHC are not the only projects that are currently discussed. Hang on to look at several of the exciting projects in the next chapter.

Notes

1. David Jackson, Maury Tigner, and Stanley Wojcicki, *The Superconducting Super-collider,* Scientific American, March 1986, page 66.
2. A first-hand account of LHC, including its history, accelerator, and experiments is given by LHC's project leader: Lyn Evans, ed., *The Large Hadron Collider, a Marvel of Technology,* EPFL Press, Lausanne, 2019. Available online from https://cds.cern.ch/record/2645935 (open access).
3. The LHC described by the CERN director general at the time of project approval: LLewellyn Smith, *The Large Hadron Collider,* Scientific American, July 2000, page 70.
4. Graham Collins, *The Discovery Machine,* Scientific American, February 2008, page 39.
5. Gordon Kane, *The Mysteries of Mass,* Scientific American, July 2005, page 40.
6. Michael Riordan, Guido Tonelli, and Sau Wu, *The Higgs at Last,* Scientific American, October 2012, page 66.
7. A first discrepancy appeared in the so-called $g - 2$ experiment at Fermilab that measured the magnetic moment of muons with unprecedented precision. Remarkably the measurement differed from the best theoretical calculations, based on the standard model, by significantly more than could be explained by tolerances. For a description, see: Daniel Garisto, *Long-Awaited Muon Measurement Boosts Evidence for New Physics,* available online from https://www.scientificamerican.com/article/long-awaited-muon-measurement-boosts-evidence-for-new-physics.
8. Supersymmetry postulates a symmetry between fermions and bosons. Every fermion has a bosonic partner and vice-versa. This new symmetry makes the theory mathematically very appealing. Unfortunately, only half the particles are known. No supersymmetric partner has shown up in experiments so far. For an early discussion, see: Gordon Kane and Howard Haber, *Is Nature Supersymmetric?,* Scientific American, June 1986, page 52.
9. Similar objectives have been pursued since the year 2000 by the *Relativistic Heavy Ion Collider* (RHIC) in Brookhaven. Two large detectors, STAR and PHENIX, observe collisions of gold atoms, accelerated to energies of 100 GeV/nucleon in two 3.8 km rings, to explore conditions immediately after the Big Bang.
10. *Luminosity* is a technical term that describes the performance of an accelerator and is very closely related to the count rate of the fundamental reactions the experimenters analyze.

12

Future Accelerators

So, what comes next? An accelerator for precision studies of the Higgs boson is high on the agenda.[1] This calls for an electron-positron machine following LHC, much like LEP did precision studies following the discoveries of Z^0 and $W^\pm$ in the Sp$\bar{\text{p}}$S. With synchrotron losses already prohibitively large in LEP, one suggestion, inspired by the SLC (Chap. 10), is to construct a linear collider. As a matter of fact, two complementary approaches are pursued, the *International Linear Collider* (ILC) and the *Compact Linear Collider* (CLIC).

Linear Colliders

The ILC,[2] a sketch with the geometry is shown in Fig. 12.1, aims to accelerate electrons (red) and positrons (blue) each to an energy of up to 250 GeV in two linear accelerators pointing at each other. Depending on the final energy, the facility will be between 30 km and 50 km long. The shorter machine would reach beam energies of 125 GeV and serve as a factory for precision measurements of the Higgs boson and its decay modes.

The ILC will work by first creating an electron beam in the source, accelerating it to a few GeV before injecting it into the electron damping ring to reduce the beam size. Once it is small enough, the electrons leave the damping ring and travel all the way to the far-end of the electron linac. From here they are accelerated to their final energy. At some point along the linac, a fraction of the electrons is directed onto a target where positrons are created. The remaining electrons follow a straight path along which an elaborate magnetic lens system reduces the beam size further before they arrive at the collision point in the detector. The positrons travel from the target to the positron damping ring in

© The Author(s), under exclusive license to Springer Nature Switzerland AG 2024
V. Ziemann, *Beams*, Copernicus Books,
https://doi.org/10.1007/978-3-031-51852-2_12

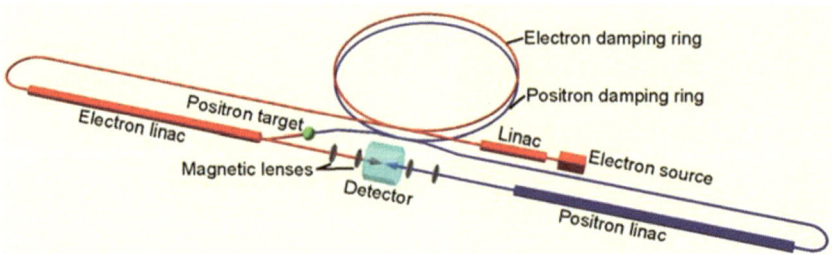

Fig. 12.1 International linear collider

order to reduce their beam size. Then they move to the far-end of the positron linac which accelerates them to their final energy. Before colliding inside the detector, their beam size is reduced with magnetic lenses as well. Like for the electrons, the sizes are reduced far below the micro-meter scale—to a size comparable to a few hundred atoms stacked on top of each other. In order to reach high collision rates, a rather large number of bunches must be accelerated. To do so efficiently, both electron and positron linacs use superconducting acceleration structures to transfer the energy to the beams.

The other linear collider project, CLIC, aims to reach beam energies up to 1500 GeV in a 50 km long facility, though shorter versions with lower energies are also considered. What energy will be realized depends to a large extent on the physics beyond the standard model that the LHC has yet to discover. The much higher beam energies in CLIC are made possible by using normalconducting acceleration structures. They can sustain much higher accelerating fields than the superconducting structures used in ILC. They can, however, sustain the high field only during a very short period of time that can be used to accelerate particles. The number of accelerated particles is therefore small and this must be compensated by much smaller beam sizes at the collision point in order to reach high collision rates. Consequently, the magnetic-lens system to reach beam sizes that are almost ten times smaller than in ILC is even more sophisticated than that for the ILC.

Despite being able to avoid the limiting effects of synchrotron radiation all linear colliders are limited by a rather moderate repetition rate of a few hundred pulses per second. In this way they are always at a disadvantage to the rings with their hundred times higher bunch-collision rates. No wonder that some colleagues came up with the *Future Circular Collider* (FCC) with a much larger circumference than LEP (Chap. 10).

Future Circular Collider

The idea behind this project is to use all existing accelerators at CERN, including the LHC, as pre-accelerators for the FCC and dig a 91 km long tunnel in the vicinity to be filled with a sequence of higher-energy accelerators.[3] Around the time LHC has completed its scientific program, sometime after 2040, the new tunnel would be home of a new electron-positron collider, called FCC-ee, that would operate with an enormous number of bunches stored in two rings that intersect inside the detectors.[4] The large circumference makes beam energies up to 200 GeV possible, only limited by restricting the power consumption to 100 MW. With its huge collision rates, FCC-ee would allow to increase the measurement precision of properties of the Z^0 and $W^\pm$, vastly exceeding the precision achieved in SLC and LEP. At beam energies around 120 GeV new ways to create Higgs bosons become accessible. And increasing the energy beyond 175 GeV makes pair production of top quarks possible. Exploring these previously unaccessible processes will allow the physicists to study tiny deviations from the standard model that lead to "new physics." FCC-ee is expected to run for at least 15 years before it will be dismantled to make space for FCC-hh.

FCC-hh is a hadron, think proton-proton, collider that replaces FCC-ee, just like LHC replaced LEP in the same tunnel. Using magnets that reach twice the field of the LHC magnets, FCC-hh would reach proton-beam energies of almost 50 TeV—seven times the energy in LHC. Like LHC it would operate with two-in-one magnets, but using the modern superconducting coils, already mentioned in the context of the high-energy LHC, to double the achievable field. The high beam energies predestine FCC-hh to find new, and heavy, particles. Moreover, high collision rates make possible searches for decay modes of particles beyond those allowed by the standard model. Beyond colliding protons, the FCC-hh would accelerate and collide beams of lead ions having beam energies of 4100 TeV. This would make it possible to probe quark-gluon plasmas with densities that were last available 10^{-12} s after the Big Bang. Assuming that the scientific lifetime of FCC-hh is at least 25 years, filling the 91 km tunnel with accelerators will keep physicists busy well into the final decades of this century.

The two versions of FCC summarize the benefits and limitations of conventional colliders very well. The electron accelerators collide point-like particles and allow precision studies. On the downside, they emit copious amounts of synchrotron radiation. At some point the cost to replenish this lost energy become prohibitive. On the other hand, proton accelerators reach much higher energies, only limited by the available magnet technology. But on the downside, protons are agglomerates of quarks and gluons, such that the collision events are very messy, making their analysis much more difficult. So, a natural question is whether there are point-like particles that are heavy. The answer is, of course, yes. Muons, the heavy brethren of the electrons, are two hundred times heavier and also point-like. Why not accelerate and collide them in a *muon collider*?

Muon Collider

The problem, however, is that muons (Chap. 3) only live for about two mil-lionth of a second. It helps, on the other hand, to accelerate them rapidly to very high energies, say 1 TeV. In that case Einstein's theory of relativity ensures that the muons' lifetime is extended to about a fiftieth of a second. That is still very short, but it gives us a fighting chance to create, accelerate, and collide muons.[5]

Just like cosmic rays impinging into the upper atmosphere create pions that decay into muons, are the muons for the collider created by smashing high-energy protons into a target to produce pions that move through a decay channel a few tens of meters long. At the other end muons come out with a large spread in energies and directions. A so-called *ionization cooling channel* needs to be developed to rapidly reduce their beam size substantially before accelerating the muons in a short linear accelerator. Then a large ring must increase their energy to a TeV in less than one millisecond. Once they reach their top energy, they are transferred to the collider ring. This ring must be as small as possible to allow the muon beams to collide at least a few thousand times before the muons decay. Much of the technology is actively being developed but requires decades of research to arrive at a convincing technical design.

All accelerators we discussed so far use metallic accelerating structures to increase the energy of the beams. The achievable fields are in all cases limited by discharges on the inner surface of the structures, which is referred to as *radio-frequency breakdown*. The high electric field literally sucks out electrons from the metal[6] and creates a plasma. The question is now whether we can use a deliberately created plasma to accelerate beams in a *plasma accelerator*.

Plasma Accelerators

Plasma accelerators are based on shooting a drive pulse, which can be either a short particle or a laser beam, into a gas. Figure 12.2 illustrates how this pulse (green) pushes the electrons (blue) in the gas away from its path and

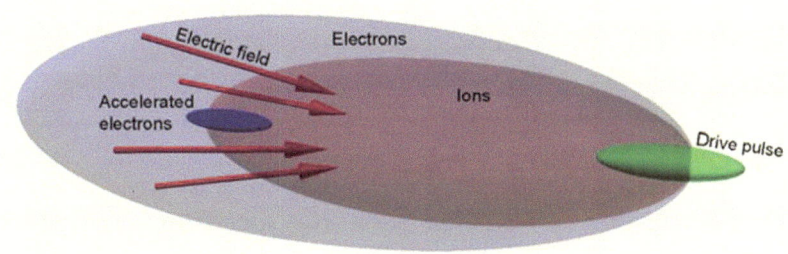

Fig. 12.2 Plasma acceleration.

thereby leaves the much more heavy and positively charged ions (red) behind. A very strong electric field (red arrows) builds up between the electrons and the ions that pulls the electrons back towards their original position such that a so-called *plasma bubble* forms. In the back of the bubble the electric field points forward, which causes electrons that are injected at just the right time to be accelerated forward.[7]

Laser-based experiments from the late 1990s onwards benefited from the invention of chirped-pulse amplification in the 1980s[8] to produce the ultra-short and high-intensity laser pulses that are needed to efficiently create plasma bubbles. In the first experiments electrons, just by chance in a good spot, were spontaneously accelerated in an uncontrolled way and came out from the plasma all over the place. After the early rather low-quality beams, over the years, the quality has improved significantly. Moreover, short bunches of electrons could be injected at just the right time into the plasma bubble. They were accelerated by several GeV over distances of centimeters. A conventional linac would require several hundred meter to reach those energies. Despite this tremendous success the energy gain in a single plasma acceleration stage is limited to a few GeV. This makes it necessary to daisy-chain multiple stages in order to reach higher energies. But this is very tricky, because the plasma bubbles are only a fraction of a millimeter short and synchronizing the electron beams in the multiple stages is very difficult. This is, however, a very active field of research with many groups at universities and national laboratories joining forces.

Instead of lasers, other groups use electron beams to create the plasma bubbles. First experiments around the turn of the millennium were based on passing the beam at the end of the 3 km linac at SLAC through a cloud of lithium vapor. Remarkably, a fraction of the electron beam had doubled its energy. This mode of operation was thus labeled "afterburner." But even in this experiment, the acceleration process was largely uncontrolled. Just a few lucky electrons were boosted to twice their original energy. In later experiments, also at SLAC in the *Facility for advanced Accelerator Experimental Tests* (FACET), a tailor-made combo of two bunches follow each other within a fraction of a millimeter and pass through a plasma. There the trailing bunch saw its energy significantly increased. Recently, that facility was upgraded and staging of multiple plasma boosters is on the agenda. A second point on the agenda is the acceleration of positrons; they are needed for plasma-based electron-positron colliders. Experiments showed that even this was possible in a prototype—so there's hope for a collider that is at least ten times shorter than ILC or CLIC, yet reaching comparable energies.

It is the limited power in the laser or the electron beam that makes staging multiple plasma cells necessary. In order to overcome this problem, the AWAKE collaboration at CERN passes a 400 GeV proton beam from the SPS (Chap. 9) through a 10 m long gas cell containing rubidium vapor. The protons need help from an external laser to form plasma bubbles, but once they do, even protons accelerate an externally injected electron beam to very high energies. Potentially they can reach the same energies as conventional linear colliders, but within a few tens of meters. Of course, an accelerator the size of the SPS is needed, but CERN already has one and AWAKE holds the promise of a TeV-scale electron accelerator, quasi as an add-on to the SPS. And even dreams of converting LHC into a plasma driver have been heard of.

But now we leave the high-energy physics thread that has pushed the development of accelerators forward and take a look at applications of accelerators in neighboring fields, where they do not only play a key role in material or life sciences, but also in medicine.

Notes

1. Keith Ellis and Beate Heinemann, *Targeting a Higgs Factory,* CERN Courier January 2021, https://cerncourier.com/a/targeting-a-higgs-factory.
2. The project leaders describe the ILC in: Barry Barish, Nicholas Walker, and Hitoshi Yamamoto, *Building the Next-Generation Collider,* Scientific American, February 2008, page 54.
3. Frank Zimmermann and Michael Benedikt, *CERN thinks bigger,* CERN Courier, June 2018 https://cerncourier.com/a/cern-thinks-bigger/.
4. In China, a comparable proposal for a *Circular Electron Positron Collider* (CEPC) is pursued. It aims to build a 100 km circumference ring to store 120 GeV beams in order to do precision measurements of the Higgs boson. It is then planned to later replace it with a *Super proton proton Collider* (SppC), much like FCC-hh follows FCC-ee. For more details see: Jie Gao, *China's bid for a Circular Electron-Positron*

Collider, CERN Courier, June 2018 https://cerncourier.com/a/chinas-bid-for-a-circular-electron-positron-collider/.

5. Daniel Schulte, *Sketching out a Muon Collider,* CERN Courier May 2020 https://cerncourier.com/a/sketching-out-a-muon-collider.

6. Actually, it is quantum-mechanical tunneling of electrons from the inside to the outside of the metal, a process referred to as Fowler-Nordheim tunneling or field emission.

7. Chandrasekhar Joshi, *Plasma Accelerators,* Scientific American, February 2006, page 40 and a more recent account https://www.scientificamerican.com/article/plasma-particle-accelerators-could-find-new-physics/.

8. Donna Strickland (b. 1959) and Gerard Mourou (b. 1944) were honored with a Nobel prize in 2018 "for groundbreaking inventions in the field of laser physics", in particular "for their method of generating high-intensity, ultra-short optical pulses," see also: Gerard Mourou and Donald Umstadter, *Extreme Light,* Scientific American, May 2002, page 80.

13

Special-Purpose Accelerators

While particle physics was the main force to drive the evolution of accelerators forward, they soon contributed to other fields of science as well. The emission of synchrotron radiation, first perceived as a nuisance, turned out to be particularly useful. A whole line-up of dedicated accelerators appeared on the scene.

Synchrotron Light Sources

As discussed in Chap. 8, synchrotron radiation was first observed in the small synchrotron in Schenectady. It took about two decades until the first dedicated synchrotron radiation source was taken into operation in 1968. After a particularly arduous construction period, the accelerator was named *Tantalus*.[1] It stored electrons with energies up to 240 MeV in a ring with a circumference of close to 10 m in order to produce visible and ultraviolet radiation.

The energy of the photons that make up the synchrotron radiation is sufficient to knock out electrons from irradiated samples. Carefully measuring the energy of the knocked-out electrons then provides insight into the atomic structure of the samples. Superconducting samples are an example from material sciences and cancer cells are another one from life sciences. This type of analysis opened up the new field of *photo emission spectroscopy* that ever since played a major role in modern light sources.[2] Appropriately, after being decommissioned in 1987, Tantalus was moved to the Archives of the Museum of National History, Smithsonian Institute in Washington, DC.

© The Author(s), under exclusive license to Springer Nature Switzerland AG 2024
V. Ziemann, *Beams*, Copernicus Books,
https://doi.org/10.1007/978-3-031-51852-2_13

Stimulated by the success of Tantalus, already during its construction, the high-energy electron ring SPEAR (Chap. 8) was equipped with a window that allowed synchrotron radiation to escape from the beam pipe. The much higher energy of the electrons in SPEAR, compared to Tantalus, produced much more energetic radiation. The photons could now knock out deep-seated electrons from inner shells of, for example, iodine, but also other materials. The energy at which deep-lying electrons become accessible is specific to each material. Irradiating with photons slightly above and then slightly below this threshold provided information about the spatial distribution of the material, for example iodine. In a pioneering experiment, today the method is called *subtraction angiography*, iodine was injected into the blood stream of a patient. After briefly exposing the patient to synchrotron radiation, his blood vessels became visible. This enables medical doctors to diagnose, among other things, clogged arteries, prone to cause heart attacks.

The energies of the emitted photons extended even into the X-ray regime with intensities much higher than available from conventional X-ray sources. The wavelengths of these photons correspond to the spacing between atoms in practically all matter. Special devices, called *monochromators* restrict the range of wavelengths, such that they become useful for so-called *diffraction experiments*. These experiments are sensitive to the crystal structure of samples,[3] even biological samples, provided they can be assembled in a crystal. In that case, the geometrical three-dimensional structure of macromolecules, such as proteins, can be determined. Today, practically all synchrotron light sources analyze thousands of proteins annually, often with important medical or biological applications. This wealth of information made the creation of a *protein data bank* necessary where all information is stored. Even the corona virus was analyzed in this way. Note how all these experiments with synchrotron light are distant relatives of the scattering experiment (Chap. 3) that Rutherford performed. Only here are photons the probes and we either look at the electrons (photoemission spectroscopy) or other photons (diffraction experiments) coming out in order to understand what happens during the scattering event.

By and by, more windows were added to SPEAR and more experiments with synchrotron light became possible. Additionally, special magnets, so-called *wigglers*[4] and *undulators*, were installed. They increased the number of emitted photons drastically by rapidly deflecting the electrons back and forth as they travel through these devices. Moreover, using permanent-magnet material made it possible to build strong-field undulators with very short periods for the deflections. Thus much higher photon energies and correspondingly shorter wavelengths became accessible. Today, practically all synchrotron light sources have multiple permanent-magnet undulators installed to generate photons in a wide range of energies for a large variety of experiments.

Retroactively the early high-energy accelerators with added synchrotron-light windows were called *first-generation light sources*. Once undulators and wigglers were added, they became *second-generation sources*. Later dedicated rings, optimized for the emission of synchrotron light, appeared. The first of these *third generation sources* was the Advanced Light Source in Berkeley that started operation in 1993. Many more appeared in the Americas, Europe, Asia, and Australia. Some of them were optimized for photon energies in the so-called soft X-ray regime that covers the important spectral region where water is transparent to radiation. This makes it possible to explore samples dissolved in water which is particularly important for biological specimen. Other machines operated at higher beam energies and were optimized for harder X-rays. They are often used to determine the structure of proteins.

These rings usually are equipped with a large number of, say 30, windows where the photons can escape the beam pipe. The photons continue through synchrotron radiation beam lines, where they are focused onto sometimes micrometer-size samples. Additionally, monochromators select a narrow range of photon energies. Exploring new materials and their properties, as well as biological or pharmaceutical materials, contributes to much of the wealth and health we enjoy today.

One of the figures of merit for light sources is the number of photons in a narrow energy range, which is called the *brightness* and determines the usability of the radiation. A huge step forward to increase it came with the invention of the *free-electron laser* (FEL).

Free-Electron Lasers

In the early 1970s, John Madey at Stanford University found that the electrons moving sideways back and forth in an undulator can exchange energy with a laser beam that also travels through the undulator.[5] This exchange works provided the beam energy, the undulator, and the laser wavelength have just the right relation with each other. By tuning the system, he could amplify the laser at the expense of the electron beam. He then placed mirrors on either side of the undulators which recirculated the light to be amplified again and again, up to rather significant power levels. It turned out that the energy of the recirculating photons fits into a very narrow range. In other words, the brightness of the laser was increased by a large factor and the radiation was suitable for diffraction experiments. This was a dream-come-true for many experimenters.

Madey's FEL worked wonderfully as long as mirrors are available, as is the case in the infra-red and visible spectral range. At higher photon energies, in the ultraviolet range, the photon energy is high enough to damage the mirrors; first they get a little opaque and later they absorb all photons. So the question came up whether free-electron lasers can be built without mirrors and, remarkably, the answer is "yes."

In the 1980s, a group in the Soviet Union and another one in Italy theoretically analyzed what happens if the electron beam current is several thousand times larger than was commonly available at the time. They found that in this case the electrons organize themselves in such a way that they copiously emit radiation in a process called *self-amplified spontaneous emission* (SASE) of radiation. The difficulty was, of course, to create the high beam currents, a problem that was solved by compressing the beams such that more electrons are packed into a smaller space. First FELs using this new technique started to emit infra-red and visible radiation in the 1990s, but towards the end of the decade a FEL at DESY in Hamburg radiated photons with energies in the far ultraviolet spectral range. Another decade later, in 2009, part of the 3 km long linear accelerator at SLAC was converted into a SASE FEL that produced X-rays millions of times brighter than was possible before. Soon a SASE FEL in Japan, named SACLA, started radiating, followed by the European XFEL at DESY in Hamburg that began its experimental program in 2017.

Determining the structure of those proteins that cannot be assembled in a crystal is the flagship experiment at these X-ray FELs. The intensity of the radiation pulse is so high that diffraction data from *single* proteins are within reach. The intensity, however, is so high that all electrons from the molecule are knocked out and the molecule explodes. Fortunately, the radiation pulse is so short, that diffraction data are collected before the molecule is blown into pieces. The analysis is not limited to macro-molecules, such as proteins, even the large-scale structure of viruses was analyzed. Instead of discussing the many exciting experiments at FELs, we turn to a second class of accelerators used to determine the structure of matter, namely *spallation neutron sources.*

Spallation Neutron Sources

Like photons, neutrons are uncharged and can therefore penetrate deeply into samples. Unlike photons, which mainly interact with the electrons, neutrons interact with the nuclei of atoms. They therefore offer a complementary view of the atomic structure of samples. Historically, neutrons, created as a by-product in nuclear power plants, were used for experiments, but they arrived as a continuous stream with only moderate intensity. In spallation sources, on the other hand, very intense proton beams are directed onto a target where they produce a short and intense pulse of neutrons in nuclear, so-called spallation, reactions. The targets are placed inside a large block of a material that slows down the neutrons to speeds of a few thousand meters per second. Their quantum-mechanical wavelength then becomes comparable to the spacing of atoms in matter, just like X-ray photons have wavelengths in this range. Since all neutrons are produced by a short pulse from the accelerator, they all have the same time of birth. After they travel some tens of meters, the higher-speed neutrons arrive early and the lower-speed neutrons come later. Selecting specific speeds or energies then becomes easy by only letting neutron within a certain time window reach the sample.

There are several large neutron spallation sources in operation. ISIS at the Rutherford-Appleton laboratory in Oxford, UK uses a ring with a circumference of 163 m. It accelerates protons to energies around 800 MeV before they are directed to targets. There the protons produce neutrons that are delivered through pipes to experimental stations. At the Paul-Scherrer Institute near Zürich, Switzerland, a large cyclotron is used to accelerate the protons. At the Spallation Neutron Source near Oak Ridge, Tennessee a superconducting linear accelerator brings the protons to 1.3 GeV before a ring "winds up" the relatively long pulses from the linac.[6] These shortened pulses are directed onto a target where they permit a high precision for the time-of-flight selection of the neutron energies. In the coming years the European Spallation Source (ESS) in Lund, Sweden, will start operating. A 500 m long superconducting linac will accelerate protons to about 2 GeV before they are directed onto the target to produce copious numbers of neutrons.

That neutrons carry a magnetic moment and behave a little like moving bar magnets makes them the perfect probe to study magnetic properties of samples, such as reducing the space needed to store one bit of information in magnetic-memory devices. Exploring how the distance between atoms changes as the temperature of samples is changed or the sample is mechanically deformed is a further field of research.

The interaction rate of neutrons with the element boron is particularly high. This is exploited by giving boron-rich medication that predominantly attaches to cancer cells in patients and exposing them to neutrons, which causes nuclear reactions mainly in the cancer cells. This form of therapy is referred to as *boron neutron capture therapy*[7] (BNCP). It is one medical application of accelerators, but there are more.

Medical Accelerators

Many larger hospitals own small electron linacs,[8] based on the same technology as the long SLAC linac from Chap. 7. The linacs are around a meter long to produce electron beams with energies in the 5–10 MeV range. These beams are either used to directly irradiate cancer cells in patients or they are directed onto targets to produce X-rays that irradiate the cells.

Even small cyclotrons are found in the basement of larger hospitals. These machines have a diameter of less than one meter and reach energies around 10 MeV with either protons or deuterons. They are mainly used to produce radioactive isotopes, used in therapy or diagnostics. An example is *positron-emission tomography*[9] (PET) where a radioactive tracer, for example, oxygen-15, is injected into a patient's blood stream. The radioactive isotopes decay and emit a characteristic radiation that pinpoints its point of origin, thus showing the position of the blood vessels.

Special medical centers—about 70 operate world-wide—own somewhat larger cyclotrons that accelerate protons up to around 230 MeV. At this energy the protons penetrate about 25 cm into water—or into the human body. They lose most of their energy at the end of their path where they deposit a high radioactive dose. This makes them particularly suitable to irradiate deep-seated tumors without unduly affecting tissue along the entrance channel or behind the tumor. A few facilities accelerate carbon ions, which has some therapeutic advantages. They are, however, based on synchrotrons and are significantly larger and more expensive to build and operate than cyclotrons.

In contrast to the accelerators for particle physics or synchrotron-radiation based research all of these smaller medical accelerators are manufactured and sold by industry. They actually represent a market worth several hundred millions of dollars annually, just like accelerators used in industry.

Industrial Accelerators

Probably the accelerators with the largest number of installations (about 10000) are *ion implanters*.[10] They alter the properties of semiconductors in a process called *doping* by using Cockroft and Walton's technology (Chap. 4) to bombard semiconductors with ions accelerated to a few 100 keV. Semiconductor manufacturers employ these machines in large numbers to produce the chips that power the computer I use to write this book.

A large number of van de Graaff accelerators (Chap. 4) are used for dating historic objects by measuring the ratio of carbon-14 to the carbon-12 in a sample. As long as organisms—often trees—live, the ratio is constant, but carbon-14 is radioactive and decays after the organism died.[11] After about 5600 years only half is left in a sample. Minute samples are sufficient to determine the age. This makes the method particularly suitable to determine the age of objects of cultural heritage found in museums. Even the Louvre in Paris owns such a small accelerator.[12]

The same type of accelerator produces a large variety of beams with energies of a few MeV. These beams are directed onto samples and whatever comes out is observed. Either particles from the beam, or the sample are detected. For example, detecting X-rays allows the experimenters to accurately characterize the sample and its properties, because many reactions are specific for a material and serve as its fingerprint.

Small electron accelerators, either linacs or betatrons, discussed in Chap. 4, smash beams of a few tens of MeV onto a target to generate X-rays. The emerging radiation is then used to sterilize objects or, allowed in some countries, to preserve food. Somewhat larger betatrons accelerate electrons up to 100 MeV and produce powerful X-rays that penetrate thick sheets of metal. They are used to image, for example, entire trucks to verify its contents without opening the doors. At even higher intensities, the weldings in large constructions, such as ships or the steel frames in buildings, are validated to make sure no bad connections are present.

One could even claim that old-fashioned thick-screen TV sets, as close descendents of a Crookes tube, are small accelerators. After all, there are 20 keV electrons guided to the target, also known as a TV screen, where they create an image for us to enjoy the latest movie, but that's probably pushing it too far.

Notes

1. In Greek mythology, Tantalus was the poor soul who was punished by standing in a pool of water without being able to drink from it.
2. For an early account, see: Ednor Rowe and John Weaver, *The Uses of Synchrotron Radiation,* Scientific American, June 1977, page 32.
3. For an overview, see Lawrence Bragg, *X-ray Crystallography,* Scientific American, July 1968, page 58. Lawrence Bragg and his father William Bragg received the Nobel prize in Physics in 1915 "for their services in the analysis of crystal structure by means of X-rays."
4. Herman Winick, *Synchrotron Radiation,* Scientific American, November 1987, page 88, and Massimo Altarelli, Fred Schlachter and Jane Cross, *Making Ultra-bright X-rays,* Scientific American, December 1998, page 66.

5. Henry Freund and Robert Parker, *Free-Electron Lasers,* Scientific American, April 1989, page 84.
6. Bill Cabage, *SNS: Neutrons for "molecular movies,"* Symmetry magazine, June 2006. Online available from https://www.symmetrymagazine.org/article/junejuly-2006/sns-neutrons-molecular-movies.
7. For a more detailed description of BNCP, see: Rolf Barth, Albert Soloway and Ralph Fairchild, *Boron Neutron Capture Therapy for Cancer,* Scientific American, October 1990, page 100.
8. David Thwaites and John Tuohy, *Back to the future: the history and development of the clinical linear accelerator,* Physics in Medicine & Biology 51 (2006) R323.
9. For an introduction to PET, see: Michel Ter-Pogossian, Marcus Raichle and Burton Sobel, *Positron Emission Tomography,* Scientific American, October 1980, page 170.
10. Frederick Morehead and Billy Crowder, *Ion Implantation,* Scientific American, April 1973, Page 64.
11. Robert Hedges and John Gowlett, *Radiocarbon Dating by Accelerator Mass Spectrometry,* Scientific American, January 1986, page 100.
12. Glenn Roberts and Kelen Tuttle, *The accelerator in the Louvre,* Symmetry magazine, May 2015. Online available from https://www.symmetrymagazine.org/article/may-2015/the-accelerator-in-the-louvre.

14

Epilogue

We've come a long way towards "figuring out the world we live in" and accelerators proved to be formidable instruments to "do the figuring". Along the way theories emerged to describe the behavior of all known particles and many additional particles were discovered. Enormous technological advances were necessary to probe the inside of atoms and then deeper into all its constituents. Many of these new technologies have profound implications for our wealth and our health.[1] Think only of the world-wide web. It was created to facilitate communication in large collaborations, and today it drives international commerce to unprecedented heights. Likewise, accelerators for medical applications help us to diagnose and treat cancer, as well as to develop new medicines with the help of synchrotron radiation.

Anyway, now, at the end of the book, it's time to summarize the fruit of all the figuring in a table of all fundamental particles and forces as we understand them today. Let's first look at the particles that constitute matter, all of them fermions.

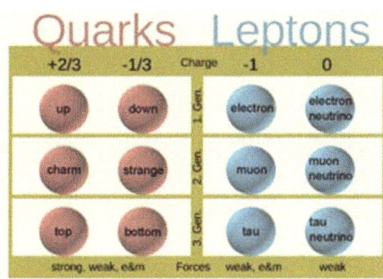

V. Ziemann, *Beams*, Copernicus Books,
https://doi.org/10.1007/978-3-031-51852-2_14

There are three generations of fundamental fermions that make up all matter. And since LEP (Chap. 10) we know that there are no further generations.

In the course of the past century, Maxwell's electromagnetic force, so-to-speak joined forces with the weak force and became the electroweak force. The nuclear force, on the other hand, has evolved from Yukawa's theory to become quantum chromodynamics or QCD and its force carriers are the gluons.

Electro-magnetic: photon

Weak: Z^0 W^+ W^-

Strong: eight x gluon

Gravity: outside standard model

And, a little on the side, there's the uncharged Higgs boson. It's purpose in life is to explain the masses of elementary particles.

Yet, despite the success of the standard model, we know that there must be more "out there" than just the standard model. It is rather unsatisfactory that many parameters of the theory, most prominently the particle masses, have to be adjusted so-to-say "by hand" rather than being constrained by some internal logic of the theory. Gravity is another worry. So far, including it in a consistent theory with the three other fundamental forces has eluded us physicists. String theory is traded as a hot candidate among certain circles, but it has so far not given us experimentally testable predictions.

And there is more! Neutrinos have the funny habit of occasionally changing from one type to another. An electron neutrino, born by a decaying neutron and after traveling some distance, all of a sudden turns into a muon neutrino or a tau neutrino. In the 1990s, this caused a puzzle, because the sun appeared to produce too few electron neutrinos to be consistent with models of the solar interior. Eventually, *neutrino oscillations*[2], as this change of neutrino identity became known, were described by a variant of Kobayashi and Maskawa's theory (Chap. 10). Several accelerators, not mentioned in the main part of this book, will explore them. I did not address them and therefore apologize to my colleagues who work there and on the many other cool accelerators that I did not mention for the sake of brevity and coherence of the story line.

Apart from accelerators, what other ways do we have to do the figuring? Only a few years ago, a new instrument to explore gravity appeared on the stage when the LIGO collaboration detected gravitational waves, already predicted by Einstein over a century ago. The source of the signal was tracked to a cataclysmic collision of black holes which are themselves corpses of super-massive stars. Many more similar events were later detected, often caused by such extreme astronomical objects as black holes or neutron stars, another end-point of stellar evolutions. Only future will tell, what else we will learn from improved versions of the LIGO detector.

Astronomy has produced many more spectacular results in recent decades, (almost) all based on catching a few rare photons with ever-improving tele-scopes. Measuring the rotation speed of stars in far-away galaxies gives us indi-cation of matter that we cannot see—we call it dark matter. Measuring the speed at which distant galaxies recede[3] tells us that there must be an unknown force that repels matter at large distances—we call it dark energy. Up to now, very little is known as to the origin of this "dark side of the theory." So, there's quite some figuring left to do!

Notes

1. An account of CERN's contribution to technological progress is given in: C. Fabjan, T. Taylor, D. Treille, and H. Wenninger, *Technology meets Research, 60 Years of CERN Technology: Selected Highlights,* World Scientific, Singapore, 2017. Available online from https://www.worldscientific.com/worldscibooks/10.1142/9921 (open access).
2. Takaaki Kajita (b. 1959) and Arthur McDonald (b. 1943) received the Nobel Prize in Physics in 2015 "for the discovery of neutrino oscillations, which shows that neutrinos have mass."
3. Saul Perlmutter (b. 1959), Brian Schmidt (b. 1967), and Adam Riess (b. 1969) received the Nobel Prize in Physics in 2011 "for the discovery of the accelerating expansion of the Universe through observations of distant supernovae."

Timeline

Time	Selected events	Chap.
before 1900	Crookes tube and X-rays	2
	Discovery of the electron and radioactivity	2
1910s	Discovery of cosmic rays	3
	Rutherford's scattering experiment	3
	Bohr's model of the atom	3
1920s	Quantum mechanics	3
	Wideroe's linear accelerator	4
1930s	Cockroft-Walton accelerator	4
	Cyclotrons	4
	Cloud chambers	3
	Discovery of positron and muon	3
1940s	Focusing of particle beams	4
	Discovery of pions	3
1950s	Cosmotron and Bevatron	5
	Antiproton and V-particles	5
	Bubble chambers	5
	The fall of parity conservation	5
1960s	CERN proton synchrotron and AGS	6
	Quark model	6
	SLAC linear accelerator	7
	Theory for the weak interaction	6

V. Ziemann, *Beams*, Copernicus Books,
https://doi.org/10.1007/978-3-031-51852-2

Time	Selected events	Chap.
1970s	Electron-positron colliders	8
	Discovery of charm quark and tau	8
	Quantum chromodynamics and gluons	8
	Fermilab main ring	9
	Discovery of bottom quark	9
	ISR and SPS at CERN	9
	Multi-wire proportional chamber	6
1980s	Proton-antiproton collider at CERN	9
	Discovery of Z^0 and $W^\pm$	9
	Tevatron at Fermilab	9
1990s	Discovery of top quark	9
	LEP at CERN and SLC at SLAC	10
	Only three generations of particles	10
2000s	B-factories	10
2010s	Large Hadron Collider	11
	Discovery of Higgs boson	11

Selected Bibliography

While writing this one, of course I dug into many other books and borrowed many good bits of information. Here is a selection of non-technical references that I found particularly useful.

The following three volumes are based on symposia where the creators of the physics and the accelerators describe their work in their own words. They are a great source of first-hand information.

- Brown L, Hoddeson L (eds) (2009) The birth of particle physics. Cambridge University Press, Cambridge, UK
- Brown L, Dresden M, Hoddeson L (eds) (2008) Pions to quarks, particle physics in the 1950s. Cambridge University Press, Cambridge, UK
- Hoddeson L, Brown L, Riordan M, Dresden M (eds) (2006) The rise of the standard model, particle physics in the 1960s and 1970s. Cambridge University Press, Cambridge, UK

The history of particle accelerators, featuring many pictures, is covered in

- Sessler A, Wilson E (2014) Engines of discovery, 2nd ed. World Scientific, Singapore

And this is a beautifully illustrated history of particle physics

- Close F, Marten M, Sutton C (2002) Particle odyssey, a journey to the heart of matter. Oxford University Press, Oxford, UK

The story around the developments leading to the discovery of the charm quark at SLAC are recounted in

- Riordan M (1987) The hunting of the quark. Simon and Schuster, New York

The early history of Lawrence's laboratory in Berkeley, the birthplace of the cyclotron, is told in

- Heilbron J, Seidel R (1989) Lawrence and his laboratory, vol I. University of California Press, Berkley. https://publishing.cdlib.org/ucpressebooks/view?docId=ft5s200764

A brief summary of the first fifty years of the German research facility DESY is available in

- 50 Years of DESY (2009) Anniversary Brochure, Deutsches Elektron Synchrotron, Hamburg. https://pr.desy.de/sites/sites_desygroups/sites_extern/site_pr/content/e104098/e104109/DESY50_anniversabrochure_E_eng.pdf

The story of Fermilab placed in its historic context is told in

- Hoddeson L, Kolb A, Westfall C (2008) Fermilab, physics, the frontier, and megascience. University of Chicago Press, Chicago

CERN dominates European particle physics since the 1950s and several books deal with its history and its impact on science and technology.

- Jungk R (1969) Die große Maschine. Deutscher Taschenbuch Verlag, München. English translation: The big machine. Charles Scribner's Sons, New York (1968)
- Kriege J (ed) (1996) History of CERN, vol III. Elsevier, Amsterdam
- Schopper H (2009) LEP—the lord of the Collider rings at CERN, 1980–2000. Springer, Heidelberg
- Evans L (ed) (2019) The large hadron collider, a marvel of technology. EPFL Press, Lausanne. https://cds.cern.ch/record/2645935/files/

- Fabjan C, Taylor T, Treille D, Wenninger H (2017) Technology meets research, 60 Years of CERN technology: selected highlights. World Scientific, Singapore. https://dx.doi.org/10.1142/9921

Another source of very readable background information over the past century is the archive of the journal *Scientific American*. As teenager in the late 1970s, I had ordered a number of reprints of classical articles from which I benefitted a lot. Access to their online archive https://www.scientificamerican.com/store/archive/ or via https://www.jstor.org should be available through many libraries. I have assembled a list of direct links to these articles at https://github.com/volkziem/Beams.

Glossary

AGS

Alternating-gradient synchrotron at BNL in the US. In operation since 1961, later converted to an injector for RHIC.

Alpha particle

Nucleus of helium atoms, composed of two protons and two neutrons.

Annihilation

The conversion of a particle and its antiparticle to photons.

Anode

The electrode that is connected to the positive pole of a voltage source.

Antiparticle

A particle with all intrinsic properties reversed to the original particle. If a particle and its antiparticle meet, they *annihilate* each other and emit photons.

Asymptotic freedom

The feature that quarks move freely inside a nucleon, but their mutual attraction, due to the color force, grows as they are pulled apart.

Baryon

Particle made of three quarks or three antiquarks. Baryons are affected by the *strong*, the *weak* and the *electro-magnetic interaction.*

Beam

Ensemble of many accelerated particles.

Brookhaven National Laboratory or BNL

Brookhaven National Laboratory on Long Island in the US, home of the Cosmotron, AGS, ISABELLE, RHIC, and the NSLS.

Boson

Particle that is not affected by the *Pauli exclusion principle*, such that multiple bosons can occupy the same state. All *force carriers* of the fundamental interactions are bosons and have integer values of the *spin* quantum number. Bosons are named after S. Bose, the Indian physicist who analyzed the behavior of this type of particle.

Bunch

An ensemble of particles that is spatially contained in three dimensions. Typically, several millions to billions of particles populate one bunch.

Calorimeter

Part of a modern particle detector to determine the energy of particles.

Cathode

The electrode that is connected to the negative pole of a voltage source.

Cathode rays

Radiation emitted from the cathode. Today we know that cathode rays are fast-traveling electrons.

CERN

The European Council for Nuclear Research in Geneva, Switzerland, home of the Proton Synchrotron (PS), ISR, SPS, Sp$\bar{\text{p}}$S, LEP, and the LHC.

Charge

The electric charge is the property of elementary particles that enables them to participate in the electro-magnetic interaction. The color charge enables them to "feel" the strong interaction, and the weak charge to "feel" the weak interaction.

Charmonium

A particle consisting of a charm and an anticharm quark.

Collider

An accelerator where counter-propagating beams collide head-on.

Color

The intrinsic property of elementary particles, in particular quarks, that enables them to "feel" the strong interaction.

Cosmotron

3 GeV proton synchrotron at BNL, operational from 1952 until 1966.

DAPHNE

Electron-positron collider in Italy, with a beam energy of 510 MeV. In operation since 1999.

DESY

Deutsches Elektronen Synchrotron in Hamburg, Germany. It is the home of several accelerators, among them the DESY synchrotron, DORIS, PETRA, HERA, FLASH, and XFEL.

Deuterium

An isotope of hydrogen. Its nucleus, called deuteron, is composed of a proton and a neutron instead of a single proton, as it is in hydrogen.

Electro-magnetic interaction

Interaction between electrically charged particles, carried by photons.

Elektroweak theory

Theory that unifies electro-magnetic and weak interactions.

Electron

The lightest charged lepton, also the particle that moves in wires and is responsible for electric currents.

Event

A collision among particles that gives rise to ejected secondary particles is called an *event*.

Fermilab

Fermi National Accelerator Laboratory near Chicago in the US. It is the home of several accelerators, among them the 400 GeV main ring and the Tevatron.

Fermion

Particle that is affected by the *Pauli exclusion principle*. It prevents two fermions of the same type to occupy the same state. All fermions have half-integer values of the *spin* quantum number. All quarks and leptons are fermions.

Feynman diagram

Graphical representation of the interaction among elementary particles.

Field theory

Theories in which the interactions among particles is effected by force-carrying bosons. Examples are QED where photons are involved, or QCD, where gluons are involved.

Fission

Splitting the nucleus of heavy atoms into multiple lighter ones.

Flavor

The six quarks, up, down, strange, charm, bottom, top are said to carry different flavors.

Fusion

Merging nuclei of light atoms to form heavier nuclei.

Gauge theory

A theory where the forces are determined by an internal symmetry. The weak, strong, and electro-magnetic forces of the standard model are based on gauge theories.

Gluon

Force-carrying boson of the strong interaction.

Gravitational interaction

Attractive force between anything with mass or energy.

Hadrons

Particles made of quarks. If three quarks are involved, they are called *baryons*. If they are made of a quark and an antiquark, they are called *mesons*. Hadrons are affected by the *strong*, the *weak* and the *electro-magnetic interaction*.

Higgs particle

Elementary particle that gives mass to all elementary particles. Discovered at CERN in 2012.

Hyperon

A baryon containing at least one strange quark. The Omega-minus is an example of a hyperon.

Interaction point

The point where two counter-propagating beams collide, usually surrounded by a detector.

Ion

An atom that has lost one or several electrons and is therefore positively charged. Some atoms can also have too many electrons to balance the charge of the nucleus and are negatively charged.

Isotope

Isotopes are nuclei with a fixed number of protons that contain different numbers of neutrons. For example, deuterium and tritium are isotopes of hydrogen with one or two additional neutrons in their respective nuclei.

ISR

Intersecting Storage Ring at CERN, a proton-proton collider that was operational from 1971 until 1984.

Jet

A directed spray of particles coming from the interaction point.

Kaon

A meson where one of the quarks is a strange or antistrange quark.

LCLS

Linac Coherent Light Source at SLAC in the US; the first X-ray free-electron laser. In operation since 2009.

LEP

Large Electron Positron collider at CERN in Geneva with a circumference of 27 km. It was in operation from 1989 until 2000 and was subsequently dismantled to make space for the LHC.

Leptons

A group of particles with the members *electron, muon, tau*, their antiparticles, and their associated neutrinos. They are only affected by the *weak* and *electromagnetic* interactions.

Linear accelerator or Linac

An accelerator to accelerate charged particles along a straight line.

LHC

The Large Hadron Collider at CERN in Geneva, which replaced *LEP* in the 27 km long tunnel. Operational since 2010.

Luminosity

The luminosity quantifies the performance of a collider in the sense that a large luminosity is equivalent to a large number of collisions. The luminosity gets bigger with increasing beam intensity and collision frequency. Moreover squeezing beams to small cross sections at the collision point increases their chance to collide and thereby the luminosity.

Mesons

Particles made of a quark and an antiquark. They are affected by the *strong*, the *weak*, and the *electro-magnetic interaction.*

Muon

A lepton very much like the electron, but about 200 times heavier.

Neutral currents

Weak interactions caused by the exchange of a Z^0 boson.

Neutrino

An uncharged lepton that only participates in the weak interaction and hardly interacts with matter at all. Each of the three fundamental leptons (electron, muon, tau) has its associated neutrino.

Neutron

The neutral partner of protons inside atomic nuclei.

Nucleon

Protons and neutrons are jointly referred to a nucleons.

Pauli exclusion principle

Prevents two same-type fermions from occupying the same state.

PEP

Positron-Electron Project at SLAC with a beam energy of 15 GeV. Operational from 1980 until 1990. Later converted into the B-factory PEP-II that operated from 1999 until 2008.

PETRA
Electron-positron collider at DESY in Hamburg with a maximum beam energy of 19 GeV.

Phi
A meson consisting of a strange and an antistrange quark.

Photon
Force-carrying boson of the *electro-magnetic interaction.*

Pion
Meson, occurring as neutral and as charged pion. Historically thought to be the carrier of the strong interaction.

Positron
The antiparticle of the electron.

Proton
The positively charged constituents of atomic nuclei. The simplest example is the atomic nucleus of hydrogen, which only consists of a single proton.

Quantum chromodynamics or QCD
Theory that describes the strong interaction among quarks and gluons.

Quantum electrodynamcis or QED
Theory that describes the electro-magnetic interactions between electrically charged particles and photons.

Quark
Fundamental constituent of hadrons, coming in six flavors: up, down, strange, charm, bottom, and top with fractional charges ($-1/3$ and $2/3$).

Radioactivity
The spontaneous decay of atomic nuclei and of elementary particles.

Resonance
Transient compounds of subatomic particles with an extremely short lifetime.

RHIC
Relativistic Heavy Ion Collider at BNL, US. In operation since 2000.

SLAC
Stanford Linear Accelerator Center in the US, home of the 3 km linear accelerator, SPEAR, PEP, the SLC, and the LCLS.

Spectrometer
Device to determine the momentum of a particle by measuring its deflection in a magnetic field.

SPEAR
Electron-positron collider at SLAC in Stanford. In operation since 1973, later converted to a synchrotron light source.

Spin

Intrinsic property of elementary particles. Particles with integer spin are called *bosons* and particles with half-integer spin are called *fermions*.

SPS

Super Proton Synchrotron at CERN. In operation since 1976.

Spp̄S

The SPS after its upgrade to collide protons and antiprotons.

SSC

Superconducting Super-Collider in the US. Terminated during construction in 1993.

Standard Model

Today's best model to describe strong, weak and electro-magnetic interactions among all known particles—quarks and leptons.

Storage ring

An accelerator to store particle beams for long times. The LHC is today's prime example.

Strong interaction

The nuclear force that binds quarks together with the help of force-carrying bosons, called *gluons*.

Superconductivity

The state in which some materials lose their electric resistance.

Supersymmetry

Theory that postulates a symmetry between fermions and bosons.

Tau

A lepton very much like the electron and muon, only very much heavier.

Tevatron

Storage ring at Fermilab to accelerate protons and later antiprotons close to one TeV.

Time projection chamber or TPC

A large gas-filled volume with superimposed electric and magnetic fields that is used to identify particle trajectories in three dimensions.

Tracker

Part of a modern detector to determine the trajectory of particles.

Transition energy

At low beam energy, increasing the particle energy mostly increases the speed of the particle. But at higher beam energies, Einstein's theory of relativity dictates that mostly the mass, rather than the speed, increases. At the transition energy, both effects balance and that has a profound influence on the stability of beams.

Tristan

Electron-positron collider at KEK in Japan with a beam energy of up to 32 GeV. Operational from 1986 until 1995. Since 1998 converted to the KEKB B-factory and 2018 upgraded to SuperKEKB.

Tritium

An isotope of hydrogen. Its nucleus is composed of a proton and two neutrons instead of a single proton, as it is in hydrogen.

UNK

Proton-proton collider in the Soviet Union. Lost funding after dissolving the Soviet Union.

Upsilon

A meson consisting of a bottom and an antibottom quark.

Vertex detector

Part of a modern detector closest to the collision point and needed to identify the position of the decay of, for example, B-mesons.

W boson

Electrically charged force carrier of the weak interaction.

Weak interaction

The force that is responsible for the *radioactivity* of particles, which is transmitted by its force carriers, the Z^0 and the $W^{\pm}$ bosons.

Z boson

Electrically neutral force carrier of the electroweak interaction.

Index

© The Editor(s) (if applicable) and The Author(s), under exclusive license to Springer Nature Switzerland AG 2024
V. Ziemann, *Beams*, Copernicus Books,
https://doi.org/10.1007/978-3-031-51852-2